Peptide Antigens

The Practical Approach Series

SERIES EDITORS

D. RICKWOOD
Department of Biology, University of Essex
Wivenhoe Park, Colchester, Essex CO4 3SQ, UK

B. D. HAMES
Department of Biochemistry and Molecular Biology
University of Leeds, Leeds LS2 9JT, UK

Affinity Chromatography
Anaerobic Microbiology
Animal Cell Culture
  (2nd Edition)
Animal Virus Pathogenesis
Antibodies I and II
Behavioural Neuroscience
Biochemical Toxicology
Biological Data Analysis
Biological Membranes
Biomechanics—Materials
Biomechanics—Structures and
  Systems
Biosensors
Carbohydrate Analysis
  (2nd Edition)
Cell–Cell Interactions
The Cell Cycle
Cell Growth and Division
Cellular Calcium
Cellular Interactions in
  Development
Cellular Neurobiology
Centrifugation (2nd Edition)

Clinical Immunology
Computers in Microbiology
Crystallization of Nucleic Acids
  and Proteins
Cytokines
The Cytoskeleton
Diagnostic Molecular Pathology
  I and II
Directed Mutagenesis
DNA Cloning I, II, and III
Drosophila
Electron Microscopy in Biology
Electron Microscopy in
  Molecular Biology
Electrophysiology
Enzyme Assays
Essential Developmental
  Biology
Essential Molecular Biology I
  and II
Experimental Neuroanatomy
Extracellular Matrix
Fermentation
Flow Cytometry (2nd Edition)

Peptide Antigens
A Practical Approach

Edited by
G. BRIAN WISDOM

Division of Biochemistry, School of Biology and Biochemistry
The Queen's University, Medical Biology Centre, Belfast, UK

OXFORD UNIVERSITY PRESS
Oxford New York Tokyo

Oxford University Press, Walton Street, Oxford OX2 6DP

Oxford   New York   Toronto
Delhi   Bombay   Calcutta   Madras   Karachi
Kuala Lumpur   Singapore   Hong Kong   Tokyo
Nairobi   Dar es Salaam   Cape Town
Melbourne   Auckland   Madrid

and associated companies in
Berlin   Ibadan

Oxford is a trade mark of Oxford University Press

A Practical Approach 🔵 is a registered trade mark
of the Chancellor, Masters, and Scholars of the University of Oxford
trading as Oxford University Press

Published in the United States
by Oxford University Press Inc., New York

A catalogue record for this book is available from the British Library

Library of Congress Cataloging in Publication Data
Peptide antigens: a practical approach/edited by G. Brian Wisdom.—1st ed.
(Practical approach series; 144)
Includes bibliographical references and index.
1. Peptides—Immunology.   2. Antigenic determinants.   3. Antigens.
I. Wisdom, G. Brian.   II. Series.
QR186.6.P76P47   1994   574.2'92–dc20   94–11639

ISBN 0 19 963452 1 (Hbk)
ISBN 0 19 963451 3 (Pbk)

Typeset by Footnote Graphics, Warminster, Wilts
Printed in Great Britain by Information Press Ltd, Eynsham, Oxon.

Preface

Peptide antigens and their antibodies have been widely exploited over many years for the measurement, location, and purification of oligopeptides such as peptide hormones. More recently, the use of peptides to mimic substructures of proteins has led to a major expansion in the application of anti-peptide antibodies. This has been especially important in identifying and characterizing proteins which are only known by their primary structure derived from a DNA sequence. Another area of recent expansion is the use of defined (or definable) peptide antigens to identify and map antibodies and T-cell receptors. Both these areas require the application of new techniques.

The aim of this volume is to provide information about the main techniques for the exploitation of peptide antigens and anti-peptide antibodies. The first chapter surveys the field and the uses of these antigens and antibodies. The remaining seven chapters describe the techniques and, in over 60 protocols, give detailed instructions for their application. The volume is to a large extent self-contained but some methods, for example monoclonal antibody production, have not been included as they are common to other areas of immunology and have already been extensively described elsewhere.

I am very grateful to my colleagues Drs Bert Rima, Neil McFerran, Brent Irvine, Pat Harriot, and David Guthrie for many useful comments during the preparation of this volume.

Belfast
September 1993

G.B.W.

Contents

Contents

7. Epitope mapping using synthetic peptides   181

Jane Worthington and Keith Morgan

8. Epitope mapping using libraries of random
peptides displayed on phage   219
William J. Dower and Steven E. Cwirla

Contributors

KAMI BEYZAVI
Bioprocessing Ltd, Medomsley Road, Consett, Co. Durham DH8 6TJ, UK.

STEVEN E. CWIRLA
Department of Molecular Biology, Affymax Research Institute, 4001 Miranda Avenue, Palo Alto, CA 94304, USA.

WILLIAM J. DOWER
Department of Molecular Biology, Affymax Research Institute, 4001 Miranda Avenue, Palo Alto, CA 94304, USA.

NIGEL P. GROOME
School of Biological and Molecular Sciences, Oxford Brooke's University, Headington, Oxford OX3 0BP, UK.

KEITH MORGAN
Department of Rheumatology, University of Manchester Medical School, Stopford Building, Oxford Road, Manchester M13 9PT, UK.

JEAN-LUC PELLEQUER
Laboratoire d'Immunochimie and Equipe de Modélisation et de Simulation des Acides Nucléiques, Institut de Biologie Moléculaire et Cellulaire du CNRS, 15 rue René Descartes, 67084 Strasbourg Cedex, France.

MICHAEL R. PRICE
Cancer Research Laboratories, Department of Pharmaceutical Sciences, University of Nottingham, Nottingham NG7 2RD, UK.

JAMES P. TAM
Department of Microbiology and Immunobiology, Vanderbilt University, Nashville, TN 37232-2363, USA.

MARC H. V. VAN REGENMORTEL
Laboratoire d'Immunochimie, Institut de Biologie Moléculaire et Cellulaire du CNRS, 15 rue René Descartes, 67084 Strasbourg Cedex, France.

BRIAN WALKER
Division of Biochemistry, School of Biology and Biochemistry, The Queen's University, Medical Biology Centre, Belfast BT9 7BL, UK.

ERIC WESTHOF
Equipe de Modélisation et de Simulation des Acides Nucléiques, Institut de Biologie Moléculaire et Cellulaire du CNRS, 15 rue René Descartes, 67084 Strasbourg Cedex, France.

Contributors

G. BRIAN WISDOM
Division of Biochemistry, School of Biology and Biochemistry, The Queen's
University, Medical Biology Centre, Belfast BT9 7BL, UK.

JANE WORTHINGTON
Arthritis and Rheumatism Council Epidemiology Research Unit, University
of Manchester Medical School, Stopford Building, Oxford Road, Manchester
M13 9PT, UK.

Abbreviations

ABTS	2,2'-azino-di-(3-ethylbenzthiazoline-6-sulphonic acid)
Acm	acetamidomethyl
ANET	Analogue Net
BDB	bis-diazotized benzidine
Boc	*t*-butyloxycarbonyl
Bom	benzyloxymethyl
BOP	benzotriazole-1-yl-oxy-tris-(dimethylamino)-phosphonium hexafluorophosphate
Br-Z	2-bromobenzoyloxycarbonyl
BSA	bovine serum albumin
BT	biotinyltyramine
Bum	butyloxymethyl
CDR	complementarity determining region
Chx	cyclohexyl
Cl-Bzl	chlorobenzyl
Cl-Z	chlorobenzoyloxycarbonyl
CPG	controlled pore glass
CS	circumsporozoite
DCC	*N,N'*-dicyclohexylcarbodiimide
DCM	dichloromethane
DIPCDI	*N,N'*-diisopropylcarbodiimide
DIPEA	diisopropylethylamine
DMA	dimethylacetamide
DMAP	4-dimethylamino pyridine
DMF	dimethylformamide
DMS	dimethyl sulphide
DMSO	dimethyl sulphoxide
Dnp	dinitrophenyl
dNTP	deoxynucleoside triphosphates (equimolar mixture of the adenine, guanine, cytosine, and thymine nucleotides)
EDC	1-ethyl-3-(3-dimethylaminopropyl)-carbodiimide
EDT	ethanedithiol
ELISA	enzyme-linked immunosorbent assay
ENET	Extension Net
fAFF	filamentous affinity phage
FMDV	foot-and-mouth disease virus
Fmoc	9-fluorenylmethyloxycarbonyl
GNET	General Net

HBTU	2-(1-*H*-benzotriazole-1-yl-)-1,1,3,3-tetramethyluronium hexafluorophosphate
HLA	human leucocyte antigen
HMMPA	4-hydroxymethyl-3-methoxyphenoxyacetic acid (resin)
HMPA	4-hydroxymethylphenoxyacetic acid (resin)
HOBt	1-hydroxybenzotriazole
HRP	horseradish peroxidase
KLH	keyhole limpet heamocyanin
mal-sac-HNSA	*N*-maleimido-6-aminocaproyl-(2′-nitro,4′-sulphonic acid)-phenyl ester
Mbh	4,4′-dimethylbenzhydryl
LE	library equivalent
MAP	multiple-antigen peptide
MBHA	*p*-methylbenzyhydrylamine (resin)
MBS	3-maleimidobenzoic acid *N*-hydroxysuccinimide ester
MeBzl	*S*-methylbenzyl
MHC	major histocompatibility complex
Mob	4-methoxybenzyl
Mtr	4-methoxy-2,3,6-trimethylbenzenesulphonyl
Mts	mesitylene-2-sulphonyl
NHS-LC-biotin	sulphosuccininimidyl-6-(bitinamido) hexanoate
NMP	*N*-methylpyrrolidinone
ON	oligonucleotide
Opfp	pentafluorophenyl
PAGE	polyacrylamide gel electrophoresis
PAL	peptide amide linker; 4-aminomethyl-3,5-dimethoxyphenoxybutyric acid
PAM	phenylacetamidomethyl (resin)
PBS	phosphate-buffered saline
PBST	phosphate-buffered saline containing Tween 20 (usually 0.05%)
PEG	polyethylene glycol
Pmc	2,2,5,7,8-pentamethylchromane-6-sulphonyl
PyBOP	benzotriazole-1-yl-oxy-tris-pyrrolidino-phosphonium hexafluorophosphate
RNET	Replacement Net
RF	replicative form
RIA	radioimmunoassay
SDS	sodium dodecyl sulphate
SPPS	solid-phase peptide synthesis
StBu	*t*-butylsulphenyl
TBS	Tris-buffered saline
tBu	*t*-butyl

TFA	triethylamine
TEAB	triethylamine-bicarbonate
TFA	trifluoroacetic acid
Tfa	trifluoroacetyl
TGFα	transforming growth factor α
TFMSA	trifluoromethane sulphonic acid
TNBSA	trinitrobenzene sulphonic acid
Tos	4-toluenesulphonyl
TMB	3,3′,5,5′-tetramethylbenzidine
Tmob	2,4,6-trimethoxybenzyl
Trt	triphenylmethane
TU	transducing units
WNET	Window Net

Peptide antigens and anti-peptide antibodies

G. BRIAN WISDOM

Antibody molecules, especially of the immunoglobulin G (IgG) class, are extremely useful reagents for the identification, measurement, purification, and characterization of various molecules (1). This arises from the fact that antibodies can be induced to most structures of molecular weight greater than 200–300 Da (provided their size is increased by attachment to a carrier) and these antibodies usually show high avidity and specificity. In addition, IgG molecules are relatively stable and can be labelled and immobilized in various ways without serious detriment to their binding abilities. Antibodies to peptide antigens have become particularly useful tools.

Methods for the accurate, rapid, and economic synthesis of peptides have become very important in many fields. The solid-phase method of Merrifield (2) has been vital for the production of defined peptide antigens and it has been developed in various ways; for example, the so-called 'tea bag' method allows the convenient and simultaneous synthesis of diverse sets of peptides for immunological purposes (3). These synthetic peptide antigens and their antibodies have had a major impact on biochemistry/molecular biology and cell biology.

1. Anti-peptide antibodies and their applications

While the measurement and subcellular location of peptides *per se* is import-ant, a major use of anti-peptide antibodies has been in the characterization of the cognate or parent protein. This use depends on the peptide mimicking one of the protein's epitopes with sufficient accuracy to allow the production of cross-reacting antibodies. The detailed investigation of antigen–antibody complexes indicates that there are commonly about 20 amino acid residues on the antigen's surface which interact with the antibody binding site and these residues are usually from different segments of the primary structure. How-ever, relatively short, linear peptides can often induce useful cross-reactive antibodies, probably as a result of induced fit (4). These antibodies are

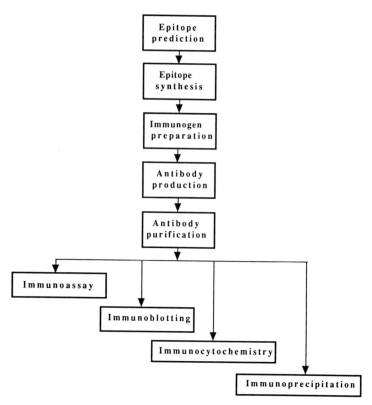

Figure 1. Sequence of steps in the development and application of anti-peptide antibodies.

especially valuable when the parent protein is rare or difficult to purify or when a probe for a specific substructure within the protein is required. Increasingly, the only information about a protein is its sequence derived from DNA, and anti-peptide antibodies provide a very important technique for further characterization of the protein.

The sequence of steps in the exploitation of peptide antigens for the characterization of proteins is shown in *Figure 1*. The prediction of antibody (or B-cell) epitopes from the primary structure has used various methods such as the identification of hydrophilic segments and turns (Chapter 2). Selected peptides (putative epitopes) can then be synthesized by either *t*-butyloxycarbonyl (Boc) or 9-fluorenylmethyloxycarbonyl (Fmoc) chemistry (Chapter 3) in amounts suitable for antibody production (about 1–5 mg). The conversion of the relatively small peptide into an immunogenic form is carried out by coupling it to a protein carrier or by synthesizing it as part of a 'multiple antigen peptide' (MAP) which has a matrix or 'tree' of three or seven lysine residues (Chapter 4). The production of anti-peptide antibodies whether as polyclonal antisera, monoclonal antibodies, or in the form of 'coliclonals' (5) is the same as for other immunogens. The antibodies in serum, ascitic fluid, etc. require purification for most applications. The IgG fraction is easily obtained, but it is often necessary to employ specific antibodies prepared by immunoaffinity chromatography (Chapter 5). The antipeptide antibodies must be carefully characterized before use. Immunoassays involving peptides have special features (Chapter 6), but the application of anti-peptide antibodies in immunoblotting, immunoprecipitation, and immunocytochemistry, is essentially the same as for other antibodies.

Anti-peptide antibodies which have demonstrated reactivity to the parent protein can be extremely valuable for its assay, purification, and location. They are also useful in characterizing protein–protein interactions, processing and other functional sites, and for determining similarities within sets of proteins. Two examples will be given to illustrate the uses of these antibodies; further examples are discussed on pp. 139–40.

Serum amyloid A apoprotein, a marker for inflammation, is difficult to purify but a sensitive and specific sandwich-type immunoassay was developed using antisera to peptides corresponding to amino acids 58–69 and 95–104 (6). The former segment was predicted to be hydrophilic and mobile and the latter peptide is the C-terminus which is lost in a processing event. The assay could thus distinguish the analyte from the degradation product.

Antibodies were raised to synthetic peptides corresponding to sequences in the α and β subunits of the insulin receptor (7). One antiserum (to residues in 1143–1154 in the regulatory region of the β subunit) immunoprecipitated the receptor only when it was phosphorylated. This result suggested that this region undergoes a conformational change when autophosphorylation occurs at one or more of the three tyrosine residues in the sequence.

An area of growing importance is the use of peptides to induce the formation of protective antibodies to pathogens. The design and evaluation of peptide vaccines is a large field and is not covered in this volume except for a discussion of the application of the MAP approach to the development of a malaria vaccine (Chapter 4). The topic is covered in more detail elsewhere (8, 9).

2. Peptide antigens and their applications

Peptide antigens may be immobilized for the immunoaffinity purification of epitope-specific antibodies (Chapter 5). Peptides are also employed extensively in immunoassays for the measurement of anti-peptide and anti-protein antibodies (Chapter 6). Increasingly, peptides are used to analyse immune responses and, in particular, for mapping the protein epitopes recognized by antibodies and T-cell receptors.

Methods used for locating epitopes on a protein include fragmenting the protein and using the fragments as probes of the antibody binding site. In the case of monoclonal antibodies a 'footprinting' technique has been described

(10); here the epitope is identified by the antibody's protective effect on fragmentation by proteolytic enzymes. Recombinant DNA methods have also been applied to the preparation of fragments of protein antigens (11); these fragments are then screened for reactivity to various antibodies.

Defined (or definable) peptides may be produced chemically or biologically in large numbers for epitope mapping. Small-scale solid-phase synthesis of peptides (of the order of 10^2–10^3) on polystyrene pins (the Pepscan approach) or on membranes (the SPOTS approach) has been widely used for mapping antibodies and, to a lesser extent, T-cell receptors (Chapter 7). An extension of this technique is the use of so-called mimotopes; these are peptides that mimic the receptor site's natural ligand. These chemical methods are not confined to the use of the 20 DNA-encoded L-amino acids and non-peptide ligands can be mimicked.

Very large numbers of random peptides (of the order of 10^8) can be produced and displayed on a filamentous phage by the introduction of oligonucleotides of random sequence in part of a phage gene coding for an exposed portion of a surface protein (Chapter 8). The phage which display peptides complementary to the antibody's binding site are selected by several rounds of an immunoaffinity procedure. The sequences of the oligonucleotide inserts in these phage are then determined, thus giving the sequence of the peptide displayed by each phage.

Numerous antibodies, including some autoantibodies, have been mapped using peptide antigens. The examination of the specificity of T-cell receptors is more complicated as this requires soluble peptide and the use of cells to present the peptide to the T-cells. Nevertheless, progress is being made rapidly (12, 13).

3. Developments

Developments in the peptide antigen field are constantly being reported. A major one is the production of synthetic peptide libraries (with 10^6–10^7 components). These are prepared by randomizing at least part of the sequence of short peptides. The resulting peptides are then screened for reactivity towards the antibody (or other receptor). In one method (14) a set of free 6-mers (with the first and second residues defined) is screened, the subset with the greatest reactivity is identified, then another set of peptides is made, with the first three residues now defined, and screened again. This process is repeated until the most reactive 6-mer is found. In another method (15) randomized 5-mers, still bound to the solid-phase resin bead, are screened and the most reactive beads are identified. Because the peptides attached to individual beads are identical, it is possible to retrieve the selected beads and subject their peptides to microsequencing. The preparation of peptide libraries has been reviewed recently (16, 17).

Of particular interest are the recent reports of the ability of peptides corresponding to the complementarity determining regions (CDRs) or hyper-variable zones of antibody polypeptides to mimic antibody binding activity. A 13-mer and a 10-mer (corresponding to the heavy-chain second CDR and the light-chain third CDR respectively) from a monoclonal antibody to lysozyme were both effective in purifying the antigen by immunoaffinity chromatography (18, 19). Peptides containing 14 to 17 residues corresponding to the the third CDR of the heavy chain of a monoclonal antibody to the human immunodeficiency virus gp160 were examined; they were shown to compete with neutralizing antibody and to be capable of inhibiting infection (20). These effects were enhanced when the peptide was cyclized; this was probably due to the peptide mimicking the loop structure of the CDR more accurately. It appears that synthetic peptide antibodies may become, like synthetic peptide antigens, a powerful tool for exploring macromolecules.

References

1. Ferrencik, M. (1993). *Handbook of immunochemistry*. Chapman and Hall, London.
2. Merrifield, R. B. (1963). *J. Am. Chem. Soc.*, **85**, 2149.
3. Houghten, R. A. (1985). *Proc. Natl. Acad. Sci. USA*, **82**, 5131.
4. Sutcliffe, J. G., Shinnick, T. M., Green, N., and Lerner, R. A. (1983). *Science*, **219**, 660.
5. Chiswell, D. J. and McCafferty, J. (1992). *Trends Biotechnol.*, **10**, 80.
6. Saile, R., Delpierre, P., Puchois, P., Hocke, G., Cachera, C., Gesquiere, J. C., Steinmetz, A., and Tartar, A. (1988). *Clin. Chem.*, **34**, 1767.
7. Perlman, R., Bottaro, D. P., White, M. F., and Kahn, C. R. (1989). *J. Biol. Chem.*, **264**, 8946.
8. Van Regenmortel, M. H. V., Briand, J. P., Muller, S., and Plaue, S. (1988). *Synthetic polypeptides as antigens*, pp. 177–91. Elsevier, Amsterdam.
9. Brown, F., Dougan, G., Hoey, E. M., Martin, S. J., Rima, B. K., and Trudgett, A. (1993). *Vaccine design*. Wiley, Chichester.
10. Sheshberadaran, H. and Payne, L. G. (1989). In *Methods in enzymology* (ed. J. J. Lagone), Vol. 178, pp. 746–64. Academic Press, London.
11. Mole, S. E. (1992). In *Methods in molecular biology* (ed. M. M. Manson), Vol. 11, pp. 105–16. Humana Press, Totowa, NJ.
12. Horsfall, A. C., Hay, F. C., Soltys, A. J., and Jones, M. G. (1991). *Immunol. Today*, **12**, 211.
13. Hay, F. C., Jones, M. G., Soltys, A., and Horsfall, A. (1993). *Immunol. Today*, **14**, 102.
14. Houghten, R. A., Pinilla, C., Blondelle, S. E., Appel, J. R., Dooley, C. T., and Cuervo, J. H. (1991). *Nature*, **354**, 84.
15. Lam, K. S., Salmon, S. E., Hersh, E. M., Hruby, V. J., Kazmeirski, W. M., and Knapp, R. J. (1991). *Nature*, **354**, 82.
16. Houghten, R. A. (1993). *Trends Genetics*, **9**, 235.

17. Houghten, R. A. (1994). *Current Biol.*, **4**, 564.
18. Welling, G. W., Geurts, T., Van Gorkum, J., Damhof, R. A., and Drijfhout, J. (1990). *J. Chromatogr.*, **512**, 337.
19. Welling, G. W., Van Gorkum, J., Damhof, R. A., Drijfhout, J. W., Bloemhoff, W., and Welling-Wester, S. (1991). *J. Chromatogr.*, **548**, 235.
20. Levi, M., Sallberg, M., Ruden, U., Herlyn, D., Maruyama, H., Wigzell, H., Marks, J., and Wahren, B. (1993). *Proc. Natl. Acad. Sci. USA*, **90**, 4374.

2

Epitope predictions from the primary structure of proteins

JEAN-LUC PELLEQUER, ERIC WESTHOF, and
MARC H. V. VAN REGENMORTEL

1. Introduction

Rapid progress in recombinant DNA techniques has led to the sequencing of an ever-increasing number of genes for which the gene products are unknown. The most common method for detecting and localizing the putative gene product is to use antibodies raised against a synthetic peptide corresponding to an epitope of the putative protein, predicted on the basis of its primary structure.

Prediction of antigenicity is thus used to indicate which peptide within the protein should be synthesized in order to obtain anti-peptide antibodies that will recognize, i.e. cross-react, with the intact protein. Such predictions rely on the existence of so-called continuous epitopes (antigenic determinants) which correspond to linear peptide fragments of the protein that are antigenically related to the parent protein. Continuous epitopes consist of a number of consecutive residues in the protein sequence and may be part of a larger so-called discontinuous epitope composed of residues distant in the sequence that are brought together by the folding of the polypeptide chain (1, 2). In the absence of further qualification, the term epitope used in this chapter refers to a B-cell epitope, i.e. an antigenic site recognized by a B-cell receptor and by an antibody molecule. A second type of epitope known as a T-cell epitope corresponds to peptide fragments of proteins that have undergone processing within an antigen-presenting cell. T-cell epitopes are recognized by T-cell receptors in association with proteins of the major histocompatibility complex (MHC) (3). The discussion of epitope prediction will be mainly limited to B-cell epitopes and only some brief reference will be made to methods developed for predicting T-cell epitopes.

No unique physico-chemical parameter has been specifically attributed to epitopes. An important property of epitopes is the accessibility needed for any recognition by B-cell receptors and by antibodies. Several methods of antigenicity prediction are therefore based on predicted accessibility. For instance, Hopp and Woods (4) postulated that since antigenic determinants should be accessible to the solvent, they should correspond to regions of the protein that are particularly hydrophilic and also contain most of the charged residues. However, statistical comparisons of various prediction algorithms using hydrophilicity scales showed that such scales were not the most successful ones for prediction of antigenicity (5).

In this chapter, we first describe the major epitope prediction methods that have been used and compare their efficacy in the case of a few well studied proteins. We then describe a new prediction method based on the prediction of turns. This prediction method was found to be particularly useful for selecting which peptides should be synthesized, since it predicted only few antigenic regions but with a high degree of confidence. In view of the labour involved in peptide synthesis and in production of cross-reactive anti-peptide antibodies, prediction of fewer antigenic peaks with a higher level of reliability represents an important advantage.

Antigenicity predictions are based on propensity scales of the 20 amino acid residues which reflect various physico-chemical parameters that have been correlated with the location of continuous epitopes in a few well characterized proteins. Parameters that have been used are hydrophilicity (4, 6), accessibility (7–9), flexibility of short segments of the polypeptide chain (10, 11), as well as various elements of secondary structure (5, 12, 13). Each scale contains 20 values, one for each amino acid. Propensity scales can be divided into the following five categories.

i. *Hydrophilicity/inverted hydrophobicity*

Hydrophilicity or inverted hydrophobicity scales are the most widely used. An inverted hydrophobicity scale corresponds to a hydrophobicity scale in which values are multiplied by −1. In hydrophobicity scales, the more hydrophobic amino acids have the highest positive values while the more hydrophilic ones have the highest negative values. In order to use these scales for antigenicity prediction, it is customary to invert them so that the more hydrophilic amino acids have the highest positive values.

Values used in hydrophilicity scales have been derived in various ways. Some of the earlier methods were based on measurement of partition coefficients between water and ethanol (14) and of the surface tension of amino acids in NaCl solution (15). A scale reflecting the hydrophobic environment of amino acids in proteins has been described by Manavalan and Ponnuswamy (16). Other approaches have relied on the measurement of water-vapour partition coefficients (17) and of the degree of exposure of residues in globular proteins (18). Some consensus scales have been described (19, 20). The commonly used hydrophilicity scales of Hopp and Woods (4) and Parker *et al.* (6) will be described in detail below.

ii. *Accessibility*

Accessibility scales are based upon the measurement of surface accessibility of each amino acid in different globular proteins. The accessible surface area of each atom is usually calculated by the method of Lee and Richards (21), either in tripeptides Gly-X-Gly or in folded proteins. The results can be expressed for instance as the proportion of residues that have 95% of their atoms buried (22).

iii. Flexibility
The flexibility scale of Karplus and Schulz (23) is based on values of the temperature factors of a few proteins of known three-dimensional structure. The temperature factor, B, is proportional to the root mean square displacement of an atom around an equilibrium position.

iv. Antigenicity
The antigenicity scale of Welling *et al.* (24) is based on the relative frequency with which amino acids are found in the epitopes of a few proteins of known antigenic structure. The occurrence of each amino acid in a series of continuous epitopes is used to construct the propensity scale.

v. Secondary structure
Elements of secondary structure such as the location of turns (12) or of helices (25) have been used to identify regions of probable antigenicity in proteins. A method based on the prediction of turns will be described below.

1.1 Construction of prediction profiles

A profile is obtained by representing the average propensity value over a small number of contiguous residues known as a window (4). Values taken from a propensity scale are assigned to each amino acid of the window. The size of the window varies depending on the scale used. For instance, a window size of five residues is well suited for a turn analysis while 11 residues is better for the analysis of amphiphilic helices. An arithmetical mean is then calculated within this window and assigned to the centre of the window (e.g. to the fourth residue in the case of a window of seven residues). The window is then shifted by one residue and the process is repeated. Each window centre is saved in an output file (*Figure 1*). Profiles can be smoothed by various procedures. Karplus and Schulz (23) for instance assigned a weight to each of the seven positions of the window in order to emphasize the central value, i.e. 0.25; 0.50; 0.75; 1.00; 0.75; 0.50; 0.25. Our PREDITOP program (13) smoothes the window by a Gaussian function in order to emphasize the centre of the window by decreasing the importance of the neighbours. The function is

$$F(x) = \frac{1}{\sigma\sqrt{2\pi}} \exp\left(\frac{-(x_i - x_j)^2}{2\sigma^2}\right)$$

where x_i is the point to be smoothed with i varying from 1 to the window length and x_j is the mean of the window (26). The best smoothing is obtained

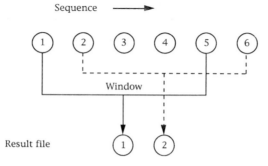

Figure 1. Window analysis for construction of prediction profiles. In this example a window of five residues is used and the arithmetical mean of the propensity value of each residue is assigned to the third residue. The window is then shifted by one residue and the mean assigned to the fourth residue. This process is repeated until the end of the sequence.

by using the value of $\sigma = 2$. In practice, each value within the window is multiplied by the corresponding value of the Gaussian function.

1.2 Comparison of prediction profiles

In order to have comparable results, all scales should be normalized. This is usually done within the values +3 and −3, which correspond to the values of the first hydrophilicity scale used for antigenicity prediction (4). The procedure will compress or expand any scale between the boundaries of ±3 by the formula: (old value × 3) / (max. old value). The results of the calculations for each successive window is a file, created in ASCII form, in which the centres of all the shifted windows are written, one per line. The first line of the result file is filled by the length of the protein and the last two lines correspond to the mean and the standard variation around the mean of all values.

Our PREDITOP (13) program calculates and displays a protein profile using the sequence of the protein and a propensity scale (hydrophilicity, hydrophobicity, accessibility, flexibility, turns, etc.). This program is described in detail elsewhere (13). *Figure 1* shows the procedure used with a window of five residues.

The output result file can be used as input for the graphical representation of the PREDITOP program and up to 10 files of the same protein can be superimposed on the screen. A colour code makes it possible to distinguish between the different predictions.

2. Common methods for predicting B-cell epitopes

Short descriptions of some widely used methods will be presented with emphasis on the methods which in a recent comparative study (5) were found to be the most reliable. A few propensity scales are presented in *Table 1*.

Table 1. Some propensity scales commonly used in antigenicity prediction

Amino acid	Hydrophilicity			TURN33 scale			
	Parker et al. (6)	Hopp and Woods (4)	Hopp (29)	Position 1	Position 2	Position 3	Position 4
R	4.2	3	0.3	−1.15	0.39	−0.05	0.19
D	10	3	2.1	3.00	−0.01	0.75	1.41
E	7.8	3	0.5	−0.92	0.24	0.21	0.28
K	5.7	3	1.4	−1.06	0.34	0.06	−0.94
S	6.5	0.3	1.8	1.34	0.98	0.25	0.46
N	7	0.2	2.3	1.65	−0.30	1.60	−0.36
Q	6	0.2	−0.2	0.11	−0.18	−0.24	0.46
G	5.7	0	3	−0.54	−0.29	3.00	2.92
P	2.1	0	2.6	0.64	3.00	−0.86	−3.00
T	5.2	−0.4	−0.1	0.27	0.34	−0.26	0.25
A	2.1	−0.5	−0.5	−0.80	−0.08	−0.38	−0.45
H	2.1	−0.5	−0.4	2.56	−0.68	0.07	−0.42
C	1.4	−1	−2.6	0.83	−0.60	−0.47	1.50
M	−4.2	−1.3	−1.8	−1.48	−0.52	−0.76	0.10
V	−3.7	−1.5	−0.7	−0.82	−0.56	−0.95	−0.33
I	−8	−1.8	−2.5	−1.52	−0.57	−0.95	−1.62
L	−9.2	−1.8	−2.5	−1.01	−0.57	−0.73	−0.94
Y	−1.9	−2.3	−2	−0.29	−0.10	0.07	0.80
F	−9.2	−2.5	−2.7	0.18	−0.37	−0.19	−0.98
W	−10	−3.4	−3	−0.96	−0.45	−0.17	0.68
α-NH2	+9.7						
α-COOH	+14.3						

i. Hydrophilicity scale of Hopp and Woods (4)

Hopp and Woods (4, 27) were the first to use the observed link between the location of epitopes at the surface of proteins and the hydrophilic character of these regions, i.e. the fact that epitopes show a high degree of exposure to the solvent. They constructed a scale for the hydrophilicity of the 20 amino acids based on the solvent parameters assigned to amino acids by Levitt (28). This scale was modified, in order to eliminate some wrong antigenic predictions found with the initial scale. For this purpose, the four charged residues (aspartic acid, glutamic acid, lysine, arginine) were given the maximum value of 3. The value for proline was increased from −1.4 to 0, as this seemed to improve antigenicity prediction within the set of 12 proteins that were studied.

Although the higher peaks in the antigenicity profiles constructed with this scale correspond to hydrophilic regions, the authors remarked that not all peaks corresponded to epitopes and that all known epitopes of a protein were not located in the most hydrophilic regions (27).

The quality of the prediction was assessed by counting the number of

known epitopes located in peaks (*C*) and outside peaks (*W*). The ratio
C/*C*+*W* was taken as the percentage of correct prediction. The accuracy of
this method will be discussed below.

ii. Acrophilicity scale (29)

In 1984 Hopp developed another method (29) for predicting surface residues
of proteins using a so-called acrophilicity scale based on his first hydrophilicity
scale as well as on the relative degree of exposure of amino acids in 49
proteins of known structure. This new scale contains the same positive values
as the hydrophilicity scale, but higher values are attributed to residues that
are frequently found at the surface of proteins (glycine, proline, asparagine,
and aspartic acid). This scale was found to be superior for locating signal and
transmembrane hydrophobic segments and was also used for predicting sites
of proteolytic processing in precursor forms of proteins as well as sites of
phosphorylation and glycosylation.

The author used stereo-paired α-carbon drawings and identified the amino
acids located in each protein protrusion. The scale was obtained by measuring
the distance between the centre of the protein and each amino acid in pro-
truding regions. It is interesting to note that the highest values were obtained
for glycine, proline, asparagine, and serine which are also the residues most
frequently found in β-turns.

iii. Antigenicity scale (24)

Welling and collaborators (24) constructed an antigenicity scale on the basis
of statistical analysis of the 606 amino acids found in 69 continuous epitopes
of 20 proteins. Each amino acid was characterized by its frequency of appearance in antigenic regions. The scale values were obtained by dividing the
frequency of appearance of an antigenic amino acid by the frequency of each
amino acid found in the National Biomedical Research Foundation protein
database (30)

This approach suffers from the imprecise nature of the antigenic database
since all residues in so-called continuous epitopes were given equal importance. It is known, however, that some residues within a continuous epitope
can be replaced by any of the other 19 amino acids without affecting antigenic
reactivity. The inclusion of such indifferent residues in the database is probably responsible for the low success rate of the antigenicity scale (5).

iv. Accessibility scale (31)

Emini and collaborators (31) who studied the VP1 of poliovirus type 1
developed an accessibility algorithm for facilitating the detection of structural
homologies revealed by the sequence alignment of hepatitis A virus and
poliovirus type 1 proteins. The method is based on the notion of surface
residue defined as a residue with $> 20 \text{ Å}^2$ of water-accessible surface found in
structural data from 28 proteins (18).

Using the fraction surface probabilities (δx) for each amino acid, a surface probability (S) at the sequence position n can be expressed by

$$S_n = \left(\prod_{i=1}^{6} \delta_{n+4-i}\right) \times (0.37)^{-6}.$$

The major difference between such a calculation and a method such as that of Hopp and Woods (4) is that the calculation procedure is based on a product instead of an addition inside the window. Although this method was not initially constructed for predicting antigenicity, we found that this algorithm predicts peaks that are well correlated with antigenicity (5).

v. Hydrophilicity scale of Parker et al. (6)

Parker and collaborators (6) constructed a new hydrophilicity scale based on peptide retention times observed during high-performance liquid chromatography (HPLC) on a reversed-phase column. The scale consists of retention times normalized from $+10$ to -10 of peptides of the type: acetyl-Gly-X-X-(Leu)$_3$-(Lys)$_2$-amide where X represents the studied amino acid. When the retention time of peptides containing charged residues was measured, it was found that the addition of a charge at the C- or N-terminus drastically affected the retention time. The scale therefore incorporated values for increased hydrophilicity brought about by the presence of a charged group at the N- and C-termini of the protein sequences. The value $+9.7$ was added to the N-terminal residue and the value $+14.3$ to the C-terminal residue.

Prediction profiles were constructed, using a window size of seven residues, by combining the surface accessibility profile of Janin (18), the flexibility profile of Karplus and Schulz (23) and their own hydrophilicity profile. From each profile, the highest peaks were given the value of 100 and the mean value was put at 0. Any residues with a profile value greater than 25% above the mean value of each profile were defined as surface sites. Then, each plot was superimposed to give the composite profile value. A 50% cut-off line in the composite profile was taken as threshold for the prediction of antigenic residues. (A program for generating composite profiles is available on diskettes for the IBM-PC or Apple Macintosh computer from the Biochemistry Department at the University of Alberta (Edmonton, Alberta, Canada T6G 2H7).)

vi. Flexibility scale (23)

In view of the observed link between antigenicity and segmental mobility (10, 11), Karplus and Schulz (23) developed a method for predicting the flexibility of protein segments. The scales are based on the known temperature factors, B, of the α-carbons of 31 proteins of known structure. The B_{norm} values were normalized following the equation $B_{norm} = (B + D)/(<B> + D)$ where $<B>$ is the average B value of all Cα atoms of the protein and D is a constant

13

developed in such a way that the root mean square deviation of the B_{norm} value is 0.3.

Amino acids could be separated into two classes comprising 10 flexible and 10 rigid residues. The rigid residues (A, L, H, V, Y, I, F, C, W, M) possessed average values lower than 1. The authors derived three scales for the amino acids called BNORM0, BNORM1, and BNORM2 corresponding to different degrees of rigidity in the neighbouring residues. BNORM0 is the scale that applies when none of the two neighbouring residues is rigid; BNORM1 is the scale applying when one of the two neighbouring residue is rigid, while BNORM2 applies when the two neighbouring residues are rigid. As a result, the prediction takes into account the flexible nature of a stretch of residues and not only the propensity of a single residue. Arithmetical weights are attributed to each of the seven positions of the window using the values 0.25; 0.5; 0.75; 1; 0.75; 0.5; 0.25.

vii. Antigenic index (32)

The first combination of different profiles each representing a different aspect of protein architecture was proposed by Jameson and Wolf (32). The chosen parameters were hydrophilicity, accessibility, flexibility, and two scales of secondary structure. After giving to each parameter a certain weight, the sum of the five curves was computed which gave rise to the so-called 'antigenic index'. The weights were chosen so that 40% of the antigenic index was derived from the secondary structure component. The accessibility and flexibility parameters, which according to these authors do not correlate well with antigenicity, were each given a weight of 15%. The remaining 30% were allocated to the inverted hydrophobicity. The choice of weights was somewhat arbitrary and it is possible that a systematic search for an optimal combination could improve the prediction success of the antigenic index.

The five scales used for computing the index are the inverted hydrophobicity (H) scale (33), the accessibility (S) scale (31), the flexibility (F) scale (23) and the secondary structure scales of Chou and Fasman (34) (CF) and Garnier *et al.* (35) (RG). Each profile was constructed with a window size of seven residues. The final curve was smoothed according to an arbitrary method compatible with the UWGCG method (32, 36). The antigenic index (A_i) is calculated according to the formula

$$A_i = \sum_{i=1}^{N} 0.3(H_i) + 0.15(S_i) + 0.15(F_i) + 0.2(CF_i) + 0.2(RG_i)$$

where H_i, S_i, F_i, CF_i, and RG_i are smoothed values obtained from the original plots of the preceding five prediction methods.

The curves obtained by this procedure display a strong bias in favour of positive values (i.e. peaks) which raises the problem of the selection of significant peaks.

3. Methods for predicting T-cell epitopes

The first author who analysed a number of peptides recognized by T-cells showed that they often possessed an amphiphilic helical structure (37). When the first crystallographic structure of a class I MHC molecule was resolved (38) it also seemed plausible that helices would fit into the putative antigen-binding pockets. Two methods have been described to locate T-cell epitopes on the basis of helix predictions. The first method displayed the periodicity of hydrophobic residues by using a so-called power spectrum procedure (39) based on a hydrophobic oscillation that repeats every θ degrees by the formula

$$P(\theta) = \sqrt{\sum_{j=k}^{k+l-1} (h_j - h_k) \sin\left(\frac{2\pi\theta}{360}\right)^2 + \sum_{j=k}^{k+l-1} (h_j - h_k) \cos\left(\frac{2\pi\theta}{360}\right)^2}$$

where h_j is the hydrophobicity of the jth residue within the kth window of length l and h_k is the average hydrophobicity of the kth window. The hydrophobic scale used in the calculation was the Fauchere–Pliska scale (39). The calculation is similar to that of the hydrophobic moment calculation (40) except that here the θ angle does not have a fixed value.

The second method determined a helix from the strip-of-helix hydrophobicity index (41). This index is the mean hydrophobicity of residues at position n, $n+4$, $n+7$, $n+11$, $n+15$. . . . This method is based on the number of turns found in helices and refers to the folding of segments of a nascent protein (42) where the hydrophobic amino acids are the nucleation residues against a hydrophobic surface such as the surface of an endosomal vesicle (43). It was found that four helical turns (15 residues) gave the best predictions of peptides presented to T-cells.

A third method of prediction of T-cell epitopes was based on a sequence pattern analysis (44). This method does not assume that T-cell epitopes are in helical structure. A sequence comparison of 57 reported T-cell epitopes highlighted a specific pattern composed of a charged residue or glycine followed by two hydrophobic residues. In several cases, the next residue was often either charged or polar. Thus the pattern is composed of four classes: charged+glycine (i.e. composed of the amino acids D, E, H, K, R, and G), hydrophobic (A, V, L, I, F, M, W, T, and Y), hydrophobic+proline (A, V, L, I, F, M, W, T, Y, and P) and polar+glycine (D, E, H, K, R, N, Q, S, T, and G). The calculation was not automated and frequency of occurrence in known T-cell epitopes was analysed.

A recent crystallographic study determined the structure of two viral peptides complexed with murine MHC class I molecules (45). Both peptides adopted an extended conformation where the N- and C-termini of the peptides were bound through extensive hydrogen bonding. It was clearly shown that there was no deep pocket that could accommodate a helical peptide.

Moreover, the epitope size was limited to only nine residues which is shorter than the helical peptides considered in earlier studies. The specificity of the binding was shown to be controlled by residues in two deep pockets called B and C in the middle of the groove (46). Peptides bound to MHC class I and II molecules were analysed by acid elution from immunoaffinity-purified MHC molecules, followed by microsequencing. Class I derived peptides had a specific length of eight to nine residues and, at certain positions, there was a strong bias for a particular amino acid (47, 48). In contrast, class II derived peptides were longer (13–24 residues) and presented no simple patterns of amino acids (49, 50). Crystal structures of MHC molecules reveal that both ends of the peptide binding groove of MHC class I molecules are closed while one end of the groove is open in the case of MHC class II molecules (51). In view of these recent findings, it is clear that existing prediction methods for T-cell epitopes will have to be reassessed.

4. Location of B-cell epitopes in proteins of known tertiary structure

Some authors have found a good correlation between the parts of a protein that protrude and the experimentally determined antigenic regions in some well studied proteins (7–9). The fact that the tertiary structure of the proteins is needed implies that the method for calculating accessibility is not a predictive method in the usual sense since it cannot be used to predict antigenicity from the primary structure.

One method starts from a basic triaxial ellipsoid with axes scaled down to half the principal axes of the protein. The molecular surfaces are then determined using a 1.4 Å probe and contoured in increments of 1 Å. The maps are displayed as Mollweide projections (7).

A second method determines the ratio between the length of the principal axes and their directions with the absolute size of the ellipsoid arbitrarily chosen to include a specified percentage of atoms. For instance, all residues which are outside the 90% ellipsoid are assigned a protrusion index of 9 (8).

Other approaches make use of the temperature factor, B, expressed along the sequence of a three-dimensional structure (10) or of measurements of the protein–water interaction energy (9). This last calculation is based on an accessible probe of 1.4 Å, but at each step a Lennard-Jones term, an electrostatic function, and a hydrogen bond function are used for energy calculation. The results are expressed as regions of high interaction energy in kcal/mol.

5. Comparison between prediction efficacies of different scales

In order to measure the efficacy of antigenicity prediction, an objective criterion must be used. It is not possible to assess the reliability of antigenicity prediction by simply counting the number of predicted peaks or by considering only the highest peak in a prediction. A reliable assessment should take into account both correct and incorrect predictions. We have used two approaches in our comparisons. The first determines the homogeneity of a prediction by calculating the difference between the number of amino acids correctly predicted and the number of amino acids incorrectly predicted (χ^2 calculation). In this case, predictions giving about the same number of amino acids inside and outside epitopes would be homogeneous and would give a low χ^2 value (5). However, a high value of χ^2 does not necessarily indicate a good prediction since it only means that there is an imbalance in the number of correct and incorrect predictions. Such a heterogeneous prediction (χ^2) may result from a large number of incorrect predictions.

The second criterion used in the comparisons simply consists of counting amino acids located inside and outside known epitopes. A threshold of 0.7 times the standard deviation around the mean is used and only amino acids above this threshold are counted. Predicted residues located in known epitopes are assigned to class A while predicted residues outside any known antigenic region are assigned to class C. The ratio A/A+C then expresses the prediction accuracy, where the ratio is 1 if there is no incorrect prediction and 0 if not a single predicted residue belongs to any known antigenic region. This ratio is more objective since it takes into account the incorrect predictions by counting residues located outside any epitope. It is also more precise since it takes into account predicted residues instead of predicted peaks.

In an earlier study the value of 22 propensity scales used for predicting antigenicity has been assessed by means of A/A+C ratios (5). The same ratio A/A+C has been used again to compute the data shown in *Table 2* which represent an updated version of our earlier results. The same 22 propensity scales were compared using 14 well studied proteins containing a total of 85 identified epitopes. Although the level of correct prediction achieved with different scales in *Table 2* is very similar to the values reported earlier (5) in a few cases the percentage of correct prediction is slightly higher. This difference is due to the fact that only proteins that have been extensively studied antigenically were included in the results of *Table 2*, whereas less well studied proteins had been included in the earlier study. It can be noted that the hydrophilicity and inverted hydrophobicity scales gave only around 57% of correct prediction. The surface accessibility scales gave lower values in both the A and C columns so that the level of correct prediction was only 53%. The acrophilicity scale of Hopp (29) which is related to surface exposure of amino acids gave results closer to the turn scales than the other accessibility scales. It is noteworthy that the flexibility scales gave the highest level of incorrect prediction and that the antigenicity scale of Welling gave the worst results, with only 44% of correct prediction (*Table 2*). We can note that the classical turn scales gave the best results with about 61% of correct prediction for the three scales. Moreover, our new turn prediction method which is based on the

Table 2. Comparative value of different prediction scales applied to 14 proteins

Parameter	Scales	Reference	Correctly predicted A	Incorrectly predicted C	Percentage of correct prediction
Inverted hydrophobicity	DOOLITTL	33	333	238	0.58
	HEIJNE	62	290	210	0.58
	MANAVALA	16	338	220	0.61
	PRILS	20	321	223	0.59
	ROSE	63	312	235	0.57
	SWEET	19	318	220	0.59
	TOTLS	20	334	213	0.61
	GES	64	271	236	0.53
	ZIMMERMA	65	284	250	0.53
Hydrophilicity	HOPP	4	271	243	0.53
	PARKER	6	311	206	0.60
Accessbility	JANIN	18	267	248	0.52
	CHOTHIA	22	289	228	0.56
	CHOTHIA8	66	239	231	0.51
	ACROPHIL	29	317	215	0.60
	EMINI	31	247	188	0.57
Flexibility	KARPLUS	23	385	383	0.50
	RAGONE	67	283	235	0.55
Antigenicity	WELLING	24	232	294	0.44
Turns	CHOUF3	34	302	186	0.62
	GARNIER3	35	330	234	0.59
	LEVITT	28	311	198	0.61
	TURN10 (addition)	52	158	146	0.52
	TURNEE (addition)	52	221	160	0.58
	TURN33 (addition)	52	204	115	0.64
	TURN10 (multiplication)	52	7	13	0.34
	TURNEE (multiplication)	52	77	33	0.70
	TURN33 (multiplication)	52	27	12	0.69

The name of each scale usually corresponds to the first eight letters of the author's name. Column A corresponds to predicted residues located in known epitopes of the 14 proteins. Column C corresponds to predicted residues located outside known antigenic regions.

The 14 proteins used were: cholera toxin, cytochrome c, surface antigen of the hepatitis B virus, human chorionic gonadotropin hormone, β-interferon, leghemoglobin, hen-egg lysozyme, myohem-erythrin, myoglobin, Gal-Gal Pyelonephritis *E. coli* pili, h-RAS p21 oncogene, human renin, scorpion neurotoxin, tobacco mosaic virus protein. The known epitopes of these proteins are listed in Pellequer *et al.* (52).

Values were obtained by the program CONTING4 contained in the PREDITOP package.

location of a four-position turn (Pellequer *et al.* (52)) gave the highest level of correct prediction (70%) of all the scales in *Table 2*.

6. Epitope prediction based on turn prediction

There is considerable evidence that antigenic sites are often located in turns of proteins. The antigenic properties of such turn peptides have been shown to resemble those of the corresponding regions in the parent protein (53–56) and a correlation between the location of turns and of continuous epitopes has been established for many proteins (57–59).

In a recent study it was shown that an algorithm used to predict the location of turns in proteins could be used to predict antigenicity (52). The basic assumption was that short peptides corresponding to turns in proteins will tend to have a turn conformation in solution. In such a case, a monoclonal antibody which recognizes a turn-like peptide should easily recognize the parent protein. This led to the development of a method in which antigenicity prediction was exclusively based on turn predictions (52). Turn scales were developed using a three-dimensional database of 87 high-resolution proteins (< 2 Å) extracted from the Brookhaven Protein Data Bank. This database contains only highly refined structures with crystallographic R-factors below 0.20 and is composed of 15 938 residues and 793 turns. The turn identification was based on the hydrogen bond pattern (60). Only four-position turns were considered, each possessing a hydrogen bond between the residue i and residue $i + 3$.

This database led to the construction of three turn scales (52). The first corresponds to helical turns, the second one to the hairpin turns, and the third one to non-specific turns. In each scale, the four positions in the turn were handled separately since a global database (mean of the frequencies in the four positions) would lead to an imprecise prediction (52). Each turn scale was normalized between the boundaries $+3$ and -3 by the PREDITOP package (13). Then, the PREDITOP program was run on the studied protein and led to four results files each corresponding to the first, second, third, and fourth position of a turn. These four files were then added or multiplied by the ADDITIO or MULTIPLI procedures included in the PREDITOP package.

The following protocol illustrates the turn prediction procedure. In all cases an ENTER value could be typed to each program query, a default value is systematically implemented except for the protein code, which is a unique word depending on the user.

Protocol 1. The turn prediction program[a]

1. Enter the protein sequence by hand or by protein database search.

2. Ascertain that the turn scales are present.

Protocol 1. *Continued*

3. Run PREDITOP.
    (a) Enter the protein code (three letter).
    (b) Enter the value 4 for simultaneous calculation (four scales).
    (c) Enter the calculation parameter: window length, centre position, smoothing procedure.
    (d) Enter scale 1, scale 2, scale 3, scale 4 (each scale corresponding to each position of a turn: TURN331, TURN332, TURN333, TURN334).

4. Use the ADDITIO procedure in order to obtain several peaks.

5. Use the MULTIPLI procedure in order to obtain few peaks.
    (a) Enter the number of files to be combined: 4.
    (b) Enter the four-result filename (saved during step **3**)
    (c) Enter the name of the unique result filename.

6. Use the graphics facilities of PREDITOP to display the prediction (e.g. *Figure 2*).

[a] The turn prediction program which includes the PREDITOP package can be obtained from the Laboratoire d'Immunochimie, IBMC, CNRS, 67084 Strasbourg Cedex, France. All requests should be accompanied by an empty diskette for IBM-PC 5¼″ or 3½″.

7. Correlation between turn prediction and antigenicity

The accuracy of antigenicity prediction was assessed by the ratio $A/A+C$. *Table 2* shows the results obtained with the helical turns scale (TURN10), the hairpin turns scale (TURNEE), and the non-specific turns scale (TURN33).

A prediction profile using the turn scale of Levitt (28) is illustrated in *Figure 2a*. An analysis of the same protein by our turn prediction method is shown in *Figure 2b*. It can be seen that our method predicts a smaller number of peaks but that no false positive peaks are obtained. Presumably the method detects the most probable turns which tend to be correlated with the major epitopes of a protein.

Table 2 lists a comparison of antigenicity predictions obtained with several scales including six turn scales. The three classical scales (CHOUF3, GARNIER3, LEVITT) gave about 61% of correct prediction while two of the three scales (TURN33, TURNEE) used in our new prediction method, with the multiplication procedure, led to the highest level of correct prediction, namely 70%. However, even when the addition procedure was used with our new method, the results were comparable with or better than classical turn predictions. The high reliability of the method is due to the fact that it attempts to predict only a limited number of epitopes, but with a high probability of success.

20

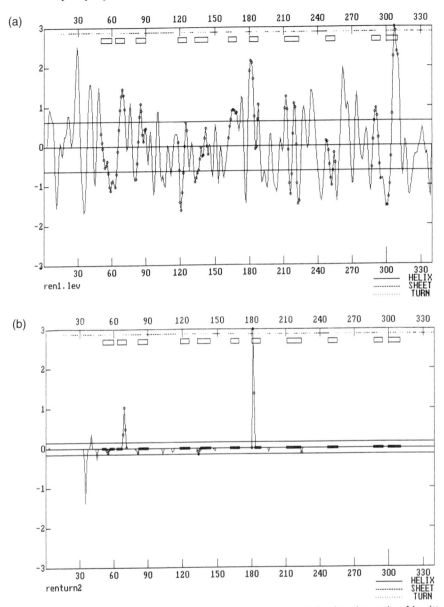

Figure 2. (a) Antigenicity prediction profile of renin constructed using the scale of Levitt (28). The graph was normalized between +3 and −3. The two full lines around the mean correspond to ±0.7 times the standard deviation. Rectangles at the top of the curves correspond to known epitopes; the circles drawn on the curves correspond to these residues. The secondary structure pattern, if known, is shown above the rectangle. The full line corresponds to a helix, a broken line corresponds to a sheet and a dotted line to a turn. (b) Antigenicity prediction profile of renin constructed using the scale TURN33 as explained in the text. Symbols are as in (a). (From Pellequer *et al.* (52).)

It is also noteworthy that the scale TURN10 is not reliable for antigenicity prediction and should be avoided. This scale which represents a particular type of turn seems not to be correlated with antigenicity.

8. Conclusions

Earlier comparisons of different prediction algorithms indicated that no single physico-chemical parameter shows a high correlation with antigenicity. It was somewhat surprising that the highest level of correct prediction calculated from the A/A+C ratio was only 60% (5) since some of the 22 scales that were compared are the most commonly used for antigenicity prediction. One reason for this fairly low level of correct prediction may be the presence in most continuous epitopes of a number of indifferent residues not directly implicated in the binding interaction with the antibody. Such replaceable residues would bias the arithmetical mean calculated on a window of five or seven residues.

The comparisons presented in *Table 2* show that the scales MANAVALA, TOTLS, PARKER, and ACROPHIL gave marginally better predictions than the other classical scales. However, the best results were obtained with the TURNEE and TURN33 scales used in conjunction with the multiplication procedure. The superiority of turn scales could be due to the fact that turns incorporate structural information which cannot be represented by a single physico-chemical parameter. Several properties of antigenic regions such as flexibility, surface accessibility, and hydrophilicity are indeed also characteristic of turns.

When antigenicity prediction is carried out to select which peptide should be synthesized to obtain anti-peptide antibodies able to cross-react with the parent protein, the turn predictions using the multiplication procedure are particularly useful. Each peak in a multiplication profile should have one residue corresponding to one turn. Then, when turns have been identified (four positions) seven residues are added to each side of the turn to constitute a 15-residue peptide. Both the TURN33 and TURNEE scales should be used since they both give a high level of correct prediction but often predict different, i.e. additional, epitopes. If a large number of antigenic regions are required, one can use the turn prediction with the addition procedure, since the score of such a prediction is still above that of most other standard prediction methods. The TURN10 scale should be avoided since helical turns did not seem to be well correlated with antigenicity.

In addition to peptides corresponding to predicted turns, it is advisable to also synthesize 15-residue long peptides corresponding to the N- and C-termini of the protein. Terminal peptides are good candidates for raising anti-peptide antibodies cross-reactive with the parent protein since chain termini are very often located at the surface of proteins (10, 61).

References

1. Berzofsky, J. A. (1985). *Science*, **229**, 932.
2. Van Regenmortel, M. H. V. (1992). In *Structure of antigens* (ed. M. H. V. Van Regenmortel), Vol. 1, p. 1. CRC Press, Boca Raton, Florida.
3. Berzofsky, J. A., Cease, K. B., Cornette, J. L., Margalit, H., Berkower, I. J., Good, M. F., Miller, L. H., and DeLisi, C. (1987). *Immunol. Rev.*, **98**, 9.
4. Hopp, T. P. and Woods, K. R. (1981). *Proc. Natl. Acad. Sci. USA*, **78**, 3824.
5. Pellequer, J. L., Westhof, E., and Van Regenmortel, M. H. V. (1991). *Methods in enzymology* (ed. J. J. Langone), Vol. 203, p. 176. Academic Press, London.
6. Parker, J. M. R., Guo, D., and Hodges, R. S. (1986). *Biochemistry*, **25**, 5425.
7. Novotny, J., Handschumacher, M., Haber, E., Bruccoleri, R. E., Carlson, W. B., Fanning, D. W., Smith, J. A., and Rose, G. D. (1986). *Proc. Natl. Acad. Sci. USA*, **83**, 226.
8. Thornton, J. M., Edwards, M. S., Taylor, W. R., and Barlow, D. J. (1986). *EMBO J.*, **5**, 409.
9. Hofmann, H.-J., Hädge, D., Höltje, M., and Höltje, H.-D. (1991). *Quant. Struct. Act. Relat.*, **10**, 300.
10. Westhof, E., Altschuh, D., Moras, D., Bloomer, A. C., Mondragon, A., Klug, A., and Van Regenmortel, M. H. V. (1984). *Nature*, **311**, 123.
11. Tainer, J. A., Getzoff, E. D., Alexander, H., Houghten, R. A., Olson, A. J., Lerner, R. A., and Hendrickson, W. A. (1984). *Nature*, **312**, 127.
12. Krchnak, V., Mach, O., and Maly, A. (1989). *Methods in enzymology* (ed. J. J. Langone), Vol. 178, p. 586. Academic Press, London.
13. Pellequer, J. L. and Westhof, E. (1993). *J. Mol. Graph.*, **11**, 204.
14. Nozaki, Y. and Tanford, C. (1971). *J. Biol. Chem.*, **246**, 2211.
15. Bull, H. B. and Breese, K. (1974). *Arch. Biochem. Biophys.*, **161**, 665.
16. Manavalan, P. and Ponnuswamy, P. K. (1978). *Nature*, **275**, 673.
17. Wolfenden, R., Andersson, L., Cullis, P. M., and Southgate, C. C. B. (1981). *Biochemistry*, **20**, 849.
18. Janin, J. (1979). *Nature*, **277**, 491.
19. Sweet, R. M. and Eisenberg, D. (1983). *J. Mol. Biol.*, **171**, 479.
20. Cornette, J. L., Cease, K. B., Margalit, H., Spouge, J. L., Berzofsky, J. A., and DeLisi, C. (1987). *J. Mol. Biol.*, **195**, 659.
21. Lee, B. and Richards, F. M. (1971). *J. Mol. Biol.*, **55**, 379.
22. Chothia, C. (1976). *J. Mol. Biol.*, **105**, 1.
23. Karplus, P. A. and Schulz, G. E. (1985). *Naturwissenschaften*, **72**, S. 212.
24. Welling, G. W., Weijer, W. J., Van der Zee, R., and Welling-Wester, S. (1985). *FEBS Lett.*, **188**, 215.
25. Hopp, T. P. (1986). *J. Immunol. Methods*, **88**, 1.
26. Van Regenmortel, M. H. V. and de Marcillac, G. D. (1988). *Immunol. Lett.*, **17**, 95.
27. Hopp, T. P. and Woods, K. R. (1983). *Mol. Immunol.*, **20**, 483.
28. Levitt, M. (1978). *Biochemistry*, **17**, 4277.
29. Hopp, T. P. (1984). *Ann. Sclavo*, **2**, 47.
30. Dayhoff, M. O. (1978). In *Atlas of protein sequence and structure*, Vol. 5. National Biomedical Research Foundation, Washington.

31. Emini, E. A., Hughes, J. V., Perlow, D. S., and Boger, J. (1985). *J. Virol.*, **55**, 836.
32. Jameson, B. A. and Wolf, H. (1988). *Comput. Applic. Biosci.*, **4**, 181.
33. Kyte, J. and Doolittle, R. F. (1982). *J. Mol. Biol.*, **157**, 105.
34. Chou, P. Y. and Fasman, G. D. (1974). *Biochemistry*, **13**, 211.
35. Garnier, J., Osguthorpe, D. J., and Robson, B. (1978). *J. Mol. Biol.*, **120**, 97.
36. Wolf, H., Modrow, S., Motz, M., Jameson, B. A., Herrmann, G., and Förtsch, B. (1988). *Comput. Applic. Biosci.*, **4**, 187.
37. DeLisi, C. and Berzofsky, J. A. (1985). *Proc. Natl. Acad. Sci. USA*, **82**, 7048.
38. Bjorkman, P. J., Saper, M. A., Samraoui, B., Bennet, W. S., Strominger, J. L., and Wiley, D. C. (1987). *Nature*, **329**, 512.
39. Margalit, H., Spouge, J. L., Cornette, J. L., Cease, K. B., Delisi, C., and Berzofsky, J. A. (1987). *J. Immunol.*, **138**, 2213.
40. Eisenberg, D., Weiss, R. M., and Terwilliger, T. C. (1984). *Proc. Natl. Acad. Sci. USA*, **81**, 140.
41. Stille, C. J., Thomas, L. J., Reyes, V. E., and Humphreys, R. E. (1987). *Mol. Immunol.*, **24**, 1021.
42. Reyes, V. E., Phillips, L., Humphreys, R. E., and Lew, R. A. (1989). *J. Biol. Chem.*, **264**, 12854.
43. Lu, S., Reyes, V. E., Bositis, C. M., Goldschmidt, T. G., Lam, V., Sorli, C. H., Torgerson, R. R., Lew, R. A., and Humphreys, R. E. (1992). In *Structure of antigens* (ed. M. H. V. Van Regenmortel), Vol. 1, p. 81. CRC Press, Boca Raton, Florida.
44. Rothbard, J. B. and Taylor, W. R. (1988). *EMBO J.*, **7**, 93.
45. Fremont, D. H., Matsumura, M., Stura, E. A., Peterson, P. A., and Wilson, I. A. (1992). *Science*, **257**, 919.
46. Matsumura, M., Fremont, D. H., Peterson, P. A., and Wilson, I. A. (1992). *Nature*, **257**, 927.
47. Falk, K., Rötzschke, O., and Rammensee, H.-G. (1990). *Nature*, **348**, 248.
48. Jardetzky, T. S., Lane, W. S., Robinson, R. A., Madden, D. R., and Wiley, D. C. (1991). *Nature*, **353**, 326.
49. Rudensky, A. Y., Preston-Hurlburt, P., Hong, S.-C., Barlow, A., and Janeway, C. A., jun. (1991). *Nature*, **353**, 622.
50. Sette, A., Ceman, S., Kubo, R. T., Sakaguchi, K., Appellea, E., Hunt, D. F., Davis, T. A., Michel, H., Shabanowitz, J., Rudersdorf, R., Grey, H. M., and DeMars, R. (1992). *Science*, **258**, 1801.
51. Brown, J. H., Jardetzky, T. S., Gorga, J. C., Stern, L. J., Urban, R. G., Strominger, J. L., and Wiley, D. C. (1993). *Nature*, **364**, 33.
52. Pellequer, J.-L., Westhof, E., and Van Regenmortel, M. H. V. (1993). *Immunol. Lett.*, **36**, 83.
53. Schulze-Gahmen, U., Prinz, H., Glatter, U., and Beyreuther, K. (1985). *EMBO J.*, **4**, 1731.
54. Rothbard, J. B., Fernandez, R., Wang, L., Teng, N. N. H., and Schoolnik, G. K. (1985). *Proc. Natl. Acad. Sci. USA*, **82**, 915.
55. Dyson, H. J., Cross, K. J., Houghten, R. A., Wilson, I. A., Wright, P. E., and Lerner, R. A. (1985). *Nature*, **318**, 480.
56. Larvor, M.-P., Djavadi-Ohaniance, L., Friguet, B., Baleux, F., and Goldberg, M. E. (1991). *Mol. Immunol.*, **28**, 523.

57. Mendz, G. L. and Moore, W. J. (1985). *Biochem. J.*, **229**, 305.
58. Williamson, M. P., Hall, M. J., and Handa, B. K. (1986). *Eur. J. Biochem.*, **158**, 527.
59. Laczko, I., Hollosi, M., Ürge, L., Ugen, K., Weiner, D. B., Mantsch, H. H., Thurin, J., and Otvös, L. J. (1992). *Biochemistry*, **31**, 4282.
60. Kabsch, W. and Sander, C. (1983). *Biopolymers*, **22**, 2577.
61. Thornton, J. M. and Sibanda, B. L. (1983). *J. Mol. Biol.*, **167**, 443.
62. Von Heijne, G. (1981). *Eur. J. Biochem.*, **116**, 419.
63. Rose, G. D., Geselowitz, A. R., Lesser, G. J., Lee, R. H., and Zehfus, M. H. (1985). *Science*, **229**, 834.
64. Engelman, D. M., Steitz, T. A., and Goldman, A. (1986). *Ann. Rev. Biophys. Biophys. Chem.*, **15**, 321.
65. Zimmerman, J. M. (1968). *J. Theor. Biol.*, **21**, 170.
66. Chothia, C. (1984). *Ann. Rev. Biochem.*, **53**, 537.
67. Ragone, R., Facchiano, F., Facchiano, A., Facchiano, A. M., and Colonna, G. (1989). *Protein Eng.*, **2**, 497.

3

Solid-phase peptide synthesis

BRIAN WALKER

1. Introduction

The chemical synthesis of oxytocin, with biological activity indistinguishable from that of the native hormone, by du Vigneaud and co-workers in 1953, represents a landmark in the application of synthetic peptides to the study of biology (1). This pioneering work earned du Vigneaud the Nobel prize for chemistry and established unequivocally the pivotal role of synthetic chemistry in peptide structure–function studies. The synthesis of this nonapeptide hormone required many months of effort and involved the combined industry of four experienced synthetic organic chemists (2). It was this labour-intensive feature that prompted Merrifield, in 1964, to introduce the solid-phase approach for the rapid production of synthetic peptides (3). This methodology has revolutionized the synthesis of peptides and has made an inestimable impact on the study of their biological activity. Fittingly, Merrifield was also awarded the Nobel prize for chemistry in 1986.

In essence, the central protocols of the original Merrifield methodology have changed very little from those detailed in the original paper, which is testament to the prescience of the instigator of the method. However, there has been a process of continual refinement in the development of protecting groups for α-amino and side-chain functionalities so as to permit the selective removal of one in the presence of the other, in the methods used to achieve complete deprotection of the final peptide, and to effect its efficient and clean removal from the inert support used in its synthesis.

In particular, the introduction of the base-labile 9-fluorenylmethyloxy-carbonyl group (Fmoc) for the protection of amino functional groups by Carpino and Han (4), laid the foundation for a genuinely orthogonal solid-phase peptide synthesis (SPPS) strategy in which the removal of the N-α-protecting group could be achieved under qualitatively different conditions required to effect cleavage of the side-chain protecting groups. This was in direct contrast to the Merrifield method which was based upon quantitative differences in the rates of acidoltyic cleavage of N-α-t-butyloxycarbonyl (Boc) groups and those protecting side-chain functionalities. Carpino's findings were subsequently exploited by Meienhoffer for the development of a polystyrene-based Fmoc SPPS strategy (5,6), which has since been brilliantly developed by R. C. Sheppard's group at the MRC laboratories in Cambridge, and which is now the synthetic method of choice for many peptide chemists throughout the world (see Section 3 for a detailed discussion).

Since the theoretical aspects of the Merrifield SPPS and Fmoc SPPS strategies have been covered in a number of excellent reviews (7–9), the aim of this chapter is to present a detailed account of the practice of peptide synthesis employing each of these methodologies; a discussion of the procedures employed for the synthesis of phosphonopeptides will be included (Section 6).

2. The Merrifield (Boc) methodology

2.1 General strategy and choice of functionalized resin

In the Merrifield approach, the peptide chain is synthesized on an inert 1% divinyl benzene cross-linked polystyrene resin via the attachment of the first Boc-protected amino acid (see *Figure 1*). Should this first amino acid, or any of the subsequent residues to be coupled, contain a nucleophilic side-chain, these will have to be masked with a protecting group that is stable to the conditions used to remove the *N*-α-Boc group (see *Figure 1*).

Peptide elaboration is achieved by removal of the Boc-protecting group from the first amino acid anchored to the support, by treatment with trifluoroacetic acid (TFA) and, after liberation of the free amino function from its trifluoroacetate salt by treatment with tertiary base, successive Boc-protected and activated amino acids are added in a series of coupling and deprotection cycles.

In essence the protocol for the synthetic elaboration of the peptide chain can be considered as a series of washing and filtration steps; these can be conveniently carried out using a manual bubbling system of the type shown in *Figure 2*. By means of the two-way tap system, N_2 can be bubbled through the reaction mixture to gently agitate the resin particles, thus ensuring a homogeneous distribution of reagents in all stages of the synthesis. Upon completion of a synthetic cycle, unreacted material, by-products, and solvent can be removed from the resin by simple filtration (through the sintered frit).

Depending on the nature of linkage (see *Figure. 1*), peptides can be obtained with a C-terminal amide (—$CONH_2$) or acid (—CO_2H) grouping. *Figure 3* lists some examples of the more commonly used linkers that are employed in the Merrifield strategy for the preparation of these two classes of peptide. The use of the classical Merrifield chloromethylated resin will yield peptide acids, as will the phenylacetamidomethyl resin (PAM). The PAM linker was introduced by Merrifield in order to circumvent the loss of peptide occurring as a result of the incomplete acid-stability of the benzyl ester bond formed between the first amino acid and the chloromethyl grouping (10). This lability can result in substantial losses of peptides containing even fewer than 10 residues; in contrast, however, the PAM linker gives a much more stable

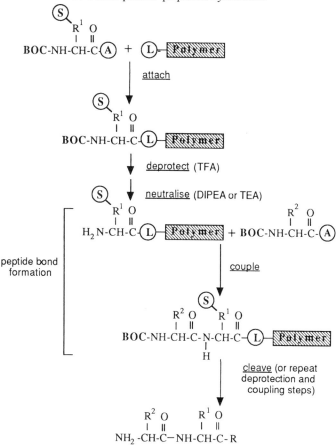

Figure 1. Flow diagram of the synthetic scheme for the solid phase synthesis of peptides using the Merrifield Boc strategy. After linking the first N-α-Boc-protected amino acid, via the carboxyl activated derivative **A**, to the resin-bound linker **L**, the peptide is elaborated by a repeated cycle of deprotection and acylation (coupling) steps. Amino acids bearing nucleophilic side-chain functional groups are incorporated using the orthogonal protecting group **S**. Depending on the method used to cleave the peptide from the resin, the product can be obtained as a C-terminal acid (R = OH), ester (R = OMe, OEt, etc.), hydrazide (R = $NHNH_2$) or amide (R = NHMe, NHEt, etc.).

ester bond and is suitable for the synthesis of longer peptides (20 residues and above).

Use of the *p*-methylbenzyhydrylamine resin (MBHA) will yield peptide amides and, since this linker is essentially completely stable to TFA, loss of peptide from the solid support due to premature acidolytic cleavage is negligible.

With each resin, cleavage of the completed peptide is usually achieved by the use of anhydrous HF or trifluoromethane sulphonic acid (TFMSA). Additionally, cleavage of the peptide from the first two resins can be achieved by nucleophilic displacement using primary amines or hydrazine to yield

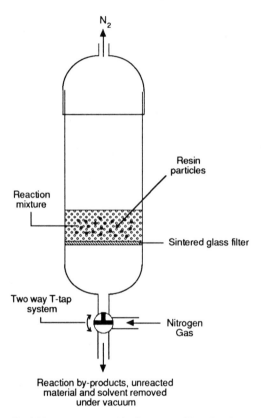

Figure 2. Diagram of bubbler system used in the manual batch-wise synthesis of peptides by conventional solid phase methods. By means of the two-way tap system, N_2 is bubbled up through the reaction mixture to agitate the resin particles. Upon completion of each synthetic cycle, by-products, unreacted material, and solvent can be removed from the resin by simple filtration, by adjusting the tap.

peptides with substituted C-terminal amides (—CONHR) or hydrazides (—CONHNH$_2$), or by transesterification with alcohol (R—OH, R = Me, Et, Pr), in the presence of tertiary base, to yield peptide esters (—COOR) (11).

The utilization of the Kaiser oxime resin, also shown in *Figure 3*, can yield peptides with a C-terminal acid or amide, depending on the strategy adopted for peptide cleavage (12).

2.2 Chemical procedures

2.2.1 Attachment of the first residue to the variously functionalized resins

i. Chloromethyl Merrifield resin

The attachment of the first Boc-amino acid to the chloromethylbenzyl resin is carried out by the nucleophilic displacement of the chloride group via a

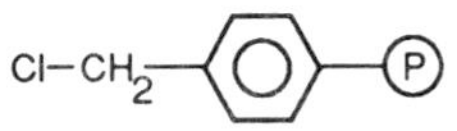

Cl—CH₂—⟨O⟩—Ⓟ

Chloromethyl (Merrifield)

HOCH₂—⟨O⟩—OCH₂CONHCH₂—⟨O⟩—Ⓟ

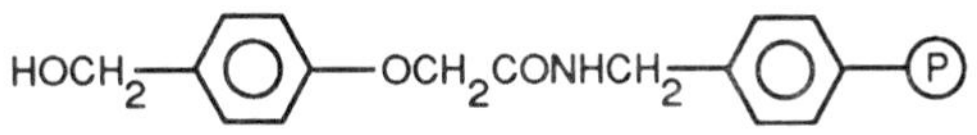

Phenylacetamidomethyl (PAM)

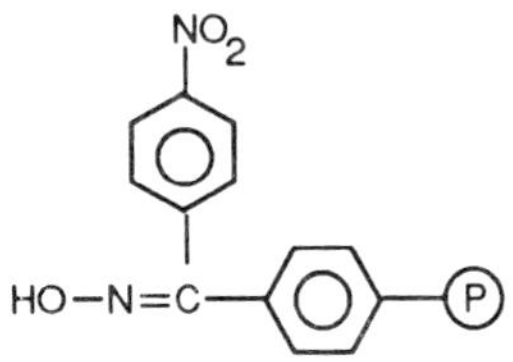

NO₂

HO—N=C—⟨O⟩—Ⓟ

p-nitrobenzophenone oxime (Kaiser oxime)

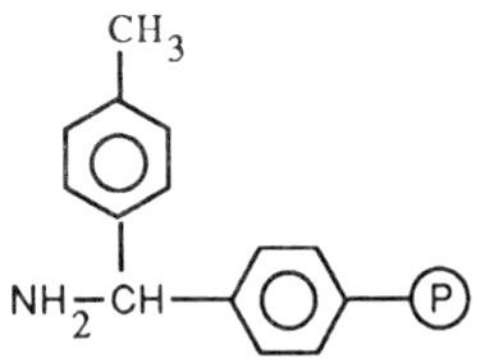

CH₃

NH₂—CH—⟨O⟩—Ⓟ

p-methylbenzhydrylamine (MBHA)

Figure 3. Structures of the most commonly used linkers employed in the synthesis of peptide acids and amides utilizing Boc SPPS.

carboxylate salt derivative of the *N*-protected amino acid. Both organic bases such as triethylamine, diisopropylethylamine, morpholine, etc. and inorganic bases such as Na_2CO_3 and Cs_2CO_3, have been used to generate the carboxylate anion. However, the caesium salt method, which was developed by Gesin (13), is reported to give the highest percentage chloride displacement and is detailed in *Protocol 1*.

Protocol 1. Attachment of the first Boc-amino acid to chloromethyl Merrifield resin

Reagents

- Boc-amino acid (available from a variety of suppliers such as Novabiochem, Peninsula Laboratories and Bachem)
- $CsCO_3$

- Dimethylformamide (DMF)
- Merrifield resin (available from a variety of suppliers such as Novabiochem, Peninsula Laboratories and Bachem)

31

Protocol 1. *Continued*

Method

1. Dissolve the Boc-amino acid (2 g) in 10–15 ml of ethanol contained in a 50–100 ml round-bottomed flask and add water (2–3 ml).

2. Adjust the pH of this solution to 7.0 (the use of narrow-range indicator paper is sufficiently accurate for this purpose), using an aqueous solution of $CsCO_3$ (~10% w/v).

3. Concentrate the solution to dryness by rotary evaporation under reduced pressure (use a Buchii Rotavap, or equivalent).

4. Add benzene (~20 ml) to the flask and again concentrate the solution to dryness by rotary evaporation. **This step must be carried out in an efficient fume cupboard as benzene is a potent carcinogen.**

5. Repeat step **4** a further four times and, in order to remove any remaining traces of water, dry the residue *in vacuo* over P_2O_5, for at least 12 h prior to use in the next step.

6. Add the dried caesium salt of the Boc-amino acid (2 mmol) to the chloromethyl Merrifield resin (~1 mmol) suspended in DMF (6–8 ml/g of resin), and gently agitate for 12 h, at 50°C. This can be carried out in the manual bubbler system shown in *Figure 2*.

7. Wash the resin in turn with the following solutions: DMF (20 ml × 5), DMF/H_2O (9:1 (v/v), 20 ml × 5), DMF (20 ml × 5), ethanol (20 ml × 5), and diethyl ether (20 ml × 5).

8. Air dry the resin before use.

Since this is a time-consuming method which does not lend itself easily to automation, a number of companies offer for sale the complete range of pre-derivatized resins containing all the naturally encoded amino acids already attached via this type of ester bond.

ii. PAM resin

The covalent attachment of the first Boc-amino acid to the PAM resin is usually achieved by reacting the symmetrical anhydride of this residue with the hydroxyl function of the linker, to form an ester bond. Since the nucleophilicity of primary alcohol is generally very low, this esterification reaction is catalysed by the presence of small amounts (0.1 mmol/1.0 mequiv of functionalized resin) of 4-dimethylamino pyridine (DMAP); however, in order to achieve maximum loading of the Boc-amino acid, it is advisable to carry out this reaction overnight (see *Protocol 2*).

Protocol 2. Attachment of the first Boc-amino acid to PAM resin

Equipment and reagents

- Manual bubbler (see *Figure 2*)
- Boc-amino acid symmetrical anhydride (*Protocol 3*)
- PAM resin (available from a variety of suppliers such as Novabiochem, Peninsula Laboratories, and Bachem)
- DMF
- Acetic anhydride
- Dichloromethane (DCM)
- 4-dimethylamino pyridine (DMAP)

Method

1. Place 1 g (1.0 mequiv) of PAM resin in the manual bubbler.[a]
2. Dissolve the symmetrical anhydride of the Boc-amino acid (4 mmol) in DMF (~20 ml/g of resin) and add this to the resin.
3. Agitate the suspension for a few minutes prior to the addition of DMAP (0.1 mmol) and then gently agitate overnight, at room temperature.
4. Wash the resin with DMF (5 × 20 ml, 3 min per wash).
5. Cap any remaining hydroxyl groups on the resin with a solution (20 ml) of acetic anhydride in DMF (10% v/v), containing DMAP (0.1 mmol). This is carried out as in step **3**.
6. Repeat step **4**.
7. Wash the resin with DCM (5 × 20 ml, 3 min per wash).
8. Air dry the resin.
9. Estimate the loading achieved by subjecting a weighed amount of the resin sample to quantitative amino acid analysis or quantitative Kaiser ninhydrin test (*Protocol 14*).

[a] Alternatively the resin can be placed in a 50 ml round-bottomed flask fitted with a magnetic stirring bar, and the reaction carried out overnight with **gentle** stirring.

Protocol 3. Preparation of Boc-amino acid symmetrical anhydrides

Reagents

- Boc-amino acid
- DCM
- DMF
- N,N'-dicyclohexylcarbodiimide (DCC) or N,N'-diisopropylcarbodiimide (DIPCDI)

Protocol 3. *Continued*

Method

1. Dissolve the Boc-amino acid (1 mmol) in 10–15 ml of freshly distilled DCM in a 50 ml round-bottomed flask fitted with a magnetic stirring bar. The addition of a few drops of DMF may be required to aid the solution of the Boc-derivatives of tryptophan, asparagine, and glutamine.

2. Add a solution of DCC (98 mg, 0.475 mmol) or DIPCDI (75 μl, 0.474 mmol) in DCM (1 ml) to the flask and stir at room temperature for 10–15 min.[a]

3. Evaporate the filtrate (or homogeneous solution in the case of the DIPCDI-mediated activation) under reduced pressure (Bucchi Rotavap, or equivalent make).

4. Dissolve the residue in DMF (~10–15 ml/1 mmol of starting amino acid). It should be used **immediately** in the coupling step.

[a] When DCC is used as the activating agent, a precipitate of dicyclohexyl urea will form and the reaction mixture should be filtered through a small sintered glass funnel (porosity No. 2 or 3) in order to remove this.

iii. MBHA resin

The attachment of the first residue to the MBHA resin can be achieved using a variety of activating agents such as N,N'-dicyclohexylcarbodiimide (DCC) or N,N'-diisopropylcarbodiimide (DIPCDI), or the more recently introduced phosphonium and uronium derivatives such as benzotriazole-l-yl-oxy-tris-(dimethylamino)-phosphonium hexafluorophosphate (BOP) (14) and 2-(1-H-benzotriazole-1-yl)-1,1,3,3-tetramethyluronium hexafluorophosphate (HBTU) (15). The carbodiimide reagents can be used for the preparation of 1-hydroxybenzotriazole (HOBt) active ester or symmetrical anhydride derivatives of the Boc amino acids (see *Figure 4*) However, the latter derivatives possess sufficient reactivity and are more frequently used; they can be prepared by the reaction of the Boc-amino acid with DCC or DIPCDI, as outlined in *Protocol 3*.

Additionally, given the favourable cost differentials between the carbodiimide reagents and HBTU or BOP, the author recommends the use of the symmetrical anhydride-mediated linkage of the first residue. In essence, this step differs little from the coupling protocols employed for the elaboration of the peptide chain discussed in Section 2.2.2.

Since the MBHA resin is usually supplied as the hydrochloride salt, it must be converted to the free base, by washing with triethylamine (TEA) or diisopropylethylamine (DIPEA), in order to couple the first residue (see *Protocol 4*).

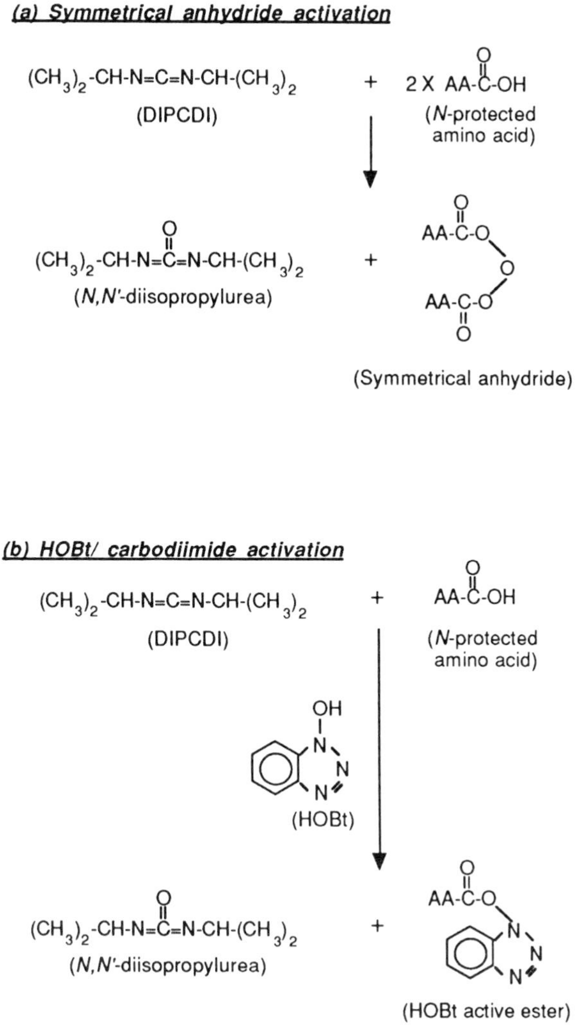

<u>**(a) Symmetrical anhydride activation**</u>

$(CH_3)_2$-CH-N=C=N-CH-$(CH_3)_2$ +   2 X AA-$\overset{\overset{\displaystyle O}{\|}}{C}$-OH

(DIPCDI)   (*N*-protected amino acid)

$(CH_3)_2$-CH-N=$\overset{\overset{\displaystyle O}{\|}}{C}$=N-CH-$(CH_3)_2$ +

(*N,N'*-diisopropylurea)

AA-$\overset{\overset{\displaystyle O}{\|}}{C}$-O$\diagdown$
          O
AA-C-O$\diagup$
     $\|$
     O

(Symmetrical anhydride)

<u>**(b) HOBt/ carbodiimide activation**</u>

$(CH_3)_2$-CH-N=C=N-CH-$(CH_3)_2$ +   AA-$\overset{\overset{\displaystyle O}{\|}}{C}$-OH

(DIPCDI)   (*N*-protected amino acid)

(HOBt)

$(CH_3)_2$-CH-N=$\overset{\overset{\displaystyle O}{\|}}{C}$=N-CH-$(CH_3)_2$ +

(*N,N'*-diisopropylurea)

(HOBt active ester)

Figure 4. Reaction schemes for the formation of symmetrical anhydride and HOBt active ester derivatives of *N*-protected amino acids.

Protocol 4. Attachment of the first Boc-amino acid to MBHA resin

Equipment and reagents

- Manual bubbler (see *Figure 2*)
- MBHA resin (available from a variety of suppliers such as Novabiochem, Peninsula Laboratories and Bachem)
- Boc-amino acid symmetrical anhydride (*Protocol 3*)

- DCM
- TEA
- DMF
- Acetic anhydride

Protocol 4. *Continued*

Method

1. Place 1 g (typically 0.3–0.6 mequiv) of MBHA resin in the manual bubbler system shown in *Figure 2.*[a]
2. Wash with DCM (5 × 20 ml, 3 min per wash).
3. Wash with 20% (v/v) TEA in DCM (2 × 20 ml, 3 min per wash).
4. Repeat step **2**.
5. Dissolve the symmetrical anhydride of the Boc-amino acid (four-fold excess over resin loading) in DMF (10–15 ml/g of resin) and add this to the resin.
6. Gently agitate the suspension for 2 h at room temperature, or until the Kaiser ninhydrin test (see *Protocol 14*) indicates that the reaction has gone to completion. If necessary, recouple by using a fresh batch of symmetrical anhydride derivative.
7. Wash with DMF (5 × 20 ml, 3 min per wash) and then with DCM (5 × 20 ml, 3 min per wash).
8. Should there be any remaining free amino groups after the repeat coupling, cap these by adding a solution (20 ml) of acetic anhydride in DMF (10% v/v) to the resin and gently agitate for 2 h at room temperature.

[a] Alternatively the resin can be placed in a 50 ml round-bottomed flask fitted with a magnetic stirring bar, and the reaction carried out overnight with **gentle** stirring.

2.2.2 Elaboration of peptide chain: deblocking, coupling, and side-chain protection

i. General considerations

The synthetic elaboration of the peptide essentially consists of the following repetitive steps:

- acidolytic removal of the Boc group from the covalently tethered first residue
- liberation of free amine from its acid salt
- coupling of the next activated Boc-amino acid

There are numerous variations as to how the first and last steps, in particular, are carried out. For example, the Boc group may be cleaved by a variety of organic acids such as TFA or formic acid (usually as solutions in DCM), mineral acids such as HCl (usually as a solution in dioxan), or Lewis acids such as $ZnCl_2$. However, the use of solutions of TFA in DCM is the generally preferred method.

Similarly, the coupling of the Boc-amino acids may be achieved using a number of activation protocols including the use of carbodiimides, phosphonium salts, or uronium derivatives.

The protocols presented below for Boc group removal and coupling of subsequent amino acids, respectively, are an amalgam of various procedures gleaned from the literature, and which the author has found to be of most general utility.

ii. Deblocking/neutralization steps

Since the acidolytic cleavage of the Boc group can generate *t*-butyl carbocations $((CH_3)_3C+)$ that are capable of alkylating residues such as cysteine, methionine, tyrosine, and tryptophan, it is essential that this repetitive deblocking step is carried out in the presence of a scavenger for these highly electrophilic species. In the author's lab we routinely use a deblocking solution of TFA in DCM (1:1, v/v) containing 2.5% anisole, (v/v) for peptides that do not contain any of the above residues. However, for peptides containing tyrosine, cysteine or methionine, the deblocking solution should also contain scavengers such as dimethyl sulphide (DMS), ethanedithiol (EDT), or thioanisole, each of which should constitute 2% (v/v) of the total solution. The author has not observed any marked difference in the effectiveness of these reagents in reducing alkylation of the above residues, but the use of DMS is preferred because it has the least offensive stench of the three. For peptides containing tryptophan, 2% dimethyl phosphite should be added to the original deblocking solution described above. Peptides containing all four of these residues should be deblocked with this latter solution **only**.

Neutralization of the TFA salt of the resin-bound amino function is then achieved by washing with a solution of DIPEA or TEA in DCM. The use of the former base is to be preferred since it gives superior results. *Protocol 5* gives a summary of the general procedure involved in Boc group removal and neutralization of resin-bound amino groups.

Protocol 5. Boc group removal and resin neutralization

Equipment and reagents

- Manual bubbler (see *Figure 2*)
- DCM
- TFA (**Caution**: see note on p. 70)
- DIPEA

Method

1. Place the Boc-amino acid- or *N*-α-Boc-peptide-derivatized resin in the manual bubbler system shown in *Figure 2*.

2. Wash with DCM (5 × 20 ml, 3 min per wash).[a]

3. Add a solution (20 ml/g of resin) of TFA in DCM (1:1, v/v)[b] to the resin and gently agitate the suspension for 30 min at room temperature.

Protocol 5. *Continued*

4. Drain the resin and wash with 20% (v/v) DIPEA in DCM (5 × 20 ml, 3 min per wash).

5. Repeat step **2**.

> *[a]* This washing step is essential in order to remove any traces of DMF that may be present, for example, following a coupling step in the synthesis. DMF forms salts with TFA which are difficult to wash out from the resin.
> *[b]* For peptides containing acid-sensitive amino acids, use an appropriate carbocation scavenger as indicated above.

iii. Activation and coupling of amino acids

The coupling of Boc-protected amino acids to the growing peptide chain can be achieved by a number of chemical procedures, and almost each peptide synthesis lab has its own preferred method. However, the use of symmetrical anhydride derivatives (see *Figure 4*), formed by reaction of two equivalents of Boc-amino acid with one equivalent of carbodiimide such as DCC or DIPCDI, as detailed in *Protocol 3*, is probably the most widely used coupling method. When DCC is used as the activating agent, the procedure followed must include a filtration step in order to remove the DCM-insoluble dicyclohexyl-urea (the by-product of the reaction), which will otherwise clog the sinter of the manual bubbler system (*Figure 2*).

The formation of symmetrical anhydride derivatives of Boc-amino acids proceeds more readily in DCM than in polar solvents such as DMF or *N*-methyl pyrrolidone, although, generally speaking, the subsequent acylation of amino functions by these derivatives occurs more rapidly in the latter solvents. This is why a solvent-exchange step, as indicated in *Protocol 3*, must be carried out before the derivative is added to the resin to effect acylation. Once the symmetrical anhydride has been formed, the coupling reaction is usually permitted to proceed for 1–2 h at room temperature. *Protocol 6* describes the general method employed for these couplings. Depending on the extent of the acylation, this step may need to be repeated using freshly prepared derivative.

Protocol 6. Symmetrical anhydride-mediated acylation of resin-bound amino groups

Reagents

- Boc-amino acid symmetrical anhydride (see *Protocol 3*)
- DCM
- DMF
- Acetic anhydride

Method

1. Remove the Boc group from the resin-bound peptide and liberate free amino groups from their TFA salts by following *Protocol 5*.

2. Prepare a four-fold excess (based on resin loading) of the symmetrical anhydride derivative of the Boc-amino acid to be coupled, according to *Protocol 3*.

3. Dissolve the derivative in DMF (10–15 ml/g of resin) and add this to the resin.

4. Gently agitate the suspension, for 1–2 h at room temperature, or until the Kaiser ninhydrin test (see *Protocol 14*) indicates that the reaction has gone to completion. If necessary, recouple using a fresh batch of the symmetrical anhydride derivative.

5. Wash with DMF (5 × 20 ml, 3 min per wash) and then with DCM (5 × 20 ml, 3 min per wash).

6. If necessary, cap any remaining amino groups by reacting with a solution (20 ml) of acetic anhydride in DMF (10% v/v), for 30–60 min at room temperature.

In addition to symmetrical anhydride-mediated condensation reactions, Boc-amino acids can also be coupled via the *in situ* formation of their 1-hydroxybenzotriazole active ester derivatives, by reaction of one equivalent of each of these reagents with a stoichiometric amount of a carbodiimide such as DCC or DIPCDI (see *Figure 4*). Additionally, the utilization of uronium- and phosphonium-mediated couplings with reagents such as HBTU (15) and BOP (14), respectively, although gaining in popularity in Boc SPPS, is more commonly associated with the Fmoc SPPS approach, and will be discussed in Section 4.

iv. Protection of side-chain functionalities

Since the formation of the peptide bond between the resin-bound amino group and the next amino acid to be coupled depends on the activation of the carboxyl group of the latter, it is imperative that the nucleophilic side-chain functional groups of the *tri*-functional amino acids be protected, otherwise branching reactions will occur. Similarly, since the unmasking of the amino group prior to each coupling step requires acidolysis with the TFA-containing solutions described above, the protecting groups chosen to mask the nucleophilic side-chain functional groups of the resin-bound amino acids such as lysine, arginine, histidine, serine, threonine, cysteine, tyrosine, aspartic acid, and glutamic acid, must possess sufficient stability to these deblocking solutions that they remain intact throughout the course of the synthesis. Premature cleavage of the side-chain protecting groups on these amino acids could result in the formation of branched peptides, with obvious dire implications for the purification of the desired product.

The most commonly used side-chain protecting groups for the Merrifield strategy are listed in *Table 1*. With the exception of the dinitrophenyl group (Dnp) used to protect the π-nitrogen of the imidazole side-chain of histidine,

Table 1. Side-chain protecting groups that are most frequently used in combination with N-α-Boc protection

Amino acid residue	Protecting group
Arginine	N-G-mesitylene-2-sulphonyl (Mts)
	N-G-4-toluenesulphonyl (Tos)
Lysine	N-ε-2-chlorobenzyloxycarbonyl (Cl-Z)
Histidine	π-2,4-dinitrophenyl (Dnp)
	π-4-toluenesulphonyl (Tos)
Cysteine	S-acetamidomethyl (Acm)
	S-methylbenzyl (MeBzl)
Tyrosine	O-2-bromobenzyloxycarbonyl (Br-Z)
Serine/threonine	O-chloro-benzyl (Cl-Bzl)
Aspartic acid/glutamic acid	O-cyclohexyl (Chx)

all of the other protecting groups shown are cleaved acidolytically using HF. The Dnp group can be removed by thiolysis using thiophenol, and this **must** be carried out **prior** to HF cleavage (see *Protocol 7*).

Protocol 7. Removal of the Dnp group from histidine-containing peptides

Equipment and reagents
- Manual bubbler system (see *Figure 2*)
- DMF
- Thiophenol
- Methanol
- Diethyl ether

Method

1. Place the resin-bound peptide in the manual bubbler device and add DMF (10 ml/g resin).

2. Gently agitate the resin suspension for 2–3 min prior to the addition of an approximately 20-fold molar excess of thiophenol (**terrible stench— must be carried out in an efficient fume cupboard**).

3. Gently agitate for a minimum of 2 h (the reaction may be left overnight without deleterious effects).

4. Drain the resin.

5. Wash with DMF (5 × 20 ml, 3 min per wash), methanol (5 × 20 ml, 3 min per wash) and finally diethyl ether (5 × 20 ml, 3 min per wash).

6. Dry *in vacuo* for 2–3 h prior to cleavage with HF, TFMSA, etc.[a]

   [a] Even after final deprotection and cleavage from the resin, the majority of peptides remain bright yellow in colour due to the presence of Dnp-thiophenol. This can be removed by gel filtration on Sephadex G-25 (Pharmacia) in 10% (v/v) acetic acid.

2.3 Cleavage of peptide from the solid support and removal of side-chain protecting groups

Because of the repetitive acidolytic cycles used to deprotect the Boc grouping prior to each coupling, the side-chain protecting groups and the peptide–resin linkage, by necessity, must possess considerable stability towards acids. The subsequent removal of these side-chain protecting groups upon completion of the synthesis and the cleavage of the peptide from the solid support, therefore entail the use of exceptionally strong acids such as HF and TFMSA. The procedures for the use of these reagents are discussed in the following two sections

i. Use of liquid HF

Removal of the side-chain protecting groups and cleavage of the peptide from the solid support can be achieved concurrently by the use of anhydrous liquid HF, **which is an exceedingly strong and corrosive acid and must be handled with extreme caution**. Herein lies, what is for some peptide chemists, the most serious drawback of the Merrifield approach. The conditions used to achieve complete deprotection and cleavage from the solid support are extremely conducive to the formation of carbocations that will destroy or chemically modify amino acid residues such as histidine, tryptophan, tyrosine, methionine, and cysteine (16). However, these potentially disastrous side reactions can be suppressed effectively if the cleavage/deprotection step is carried out in the presence of various carbocation scavengers, the choice of which depends on the nature of the amino acids contained within the peptide sequence. The damage to these sensitive residues can be further suppressed by the use of the so called 'Low–High' HF cleavage procedure (17, 18). In this method, the side-chain protecting groups are removed by using a low concentration of HF in the presence of a large molar excess of scavenger(s) which effectively compete with the sensitive amino acid residues for the carbocations. The peptide is then cleaved from the solid support, in a second step, employing a high concentration of HF.

Protocol 8 gives the experimental procedure for the standard HF cleavage, whereas *Protocol 9* details the essential features of the 'Low–High' HF procedure. **It cannot be stressed too strongly that this acid is extremely dangerous to use and special apparatus of the type shown in Figure 5 needs to be employed for the safe handling of this reagent.** This figure is included only to give an impression of the general form that an 'HF-line' takes; there are many variations on this theme. However, most will contain a condensing chamber, into which the gaseous HF is liquefied by cooling, and into which is normally placed CoF_3 (which serves as a drying agent for the gas), and a reaction vessel, into which is placed the resin-bound peptide to be cleaved. The apparatus is constructed from machined polytetrafluoroethylene (Teflon) or other such HF-resistant polymers; **under no circumstances is HF to be allowed to come**

41

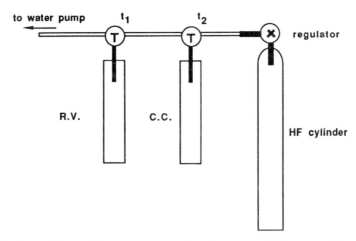

Figure 5. General form of the apparatus used to deprotect and cleave peptides with HF. The gaseous HF is admitted, by means of the tap **t**, to the condensing chamber (**C.C.**), maintained at −5 to 0 °C. It is then admitted to the evacuated reaction vessel (**R.V.**), containing the peptide resin. After the requisite reaction period (see text), the HF is removed under reduced pressure. The entire apparatus is manufactured from machined Teflon.

into contact with glass since it reacts with this only slightly less violently than it does with water! Both the reaction vessel and condensing chamber have machined-threaded tops which can be screwed securely into position with the rest of the HF line.

No detailed protocol will be presented for the use of this typical HF line, as each manufacturer will have their own carefully evaluated protocols for the use of individual apparatus and it is essential that these be followed exactly. **On no account is an HF apparatus to be used without proper instruction from a qualified operator. It is imperative that the HF line be placed in an efficient fume cupboard preferably one dedicated to this use only. The use of protective clothing including rubber gloves—preferably gauntlets—and safety visor (full face) is to be considered mandatory throughout the entire cleavage procedure.**

Protocol 8. Standard HF cleavage (0.1–0.2 mmol scale)

Equipment and reagents

- HF apparatus (the author uses an apparatus obtained from Peninsula Laboratories)
- Anisole/DMS (1:1, v/v)
- Diethyl ether or methyl-*t*-butyl ether
- Methyl butyl ether

Method

1. Place the resin-bound peptide and magnetic stirring bar (**Teflon-coated!**) into the reaction vessel of the HF apparatus.

---

2. Add the carbocation scavenger mixture, anisole/DMS, to the resin (~2.2 ml). For peptides containing cysteine residues this should be composed of anisole/DMS/p-thiocresol (1:1:0.2, v/v).

3. Replace the reaction vessel in the HF line and cool it in a dry ice/methanol bath for at least 10–15 min before admitting the HF.

4. Distil about 10 ml of HF into the reaction vessel by **following** the manufacturer's instructions **exactly** as indicated.

5. Maintain the temperature of the reaction between −5°C and 0°C for 1 h for peptides not containing Arg(Tos), otherwise the cleavage reaction will require 2 h.

6. Remove all traces of HF from the reaction vessel by evaporation under reduced pressure (follow the manufacturer's instructions).

7. Wash the resin with diethyl ether (2 × 15 ml), allow the resin and peptide to settle (most peptides are insoluble in this solvent), and then carefully decant the solvent.

8. Suspend the resin and peptide in TFA (10–15 ml) and gently agitate by stirring for a few minutes.

9. Transfer the contents of the flask, with the aid of a wide-bore Pasteur pipette, to a sintered glass funnel (No. 1 or 2 porosity) that has been fitted with a round-bottomed flask charged with about 25 ml of diethyl ether or methyl t-butyl ether.[a]

10. Wash out the contents of the round-bottomed flask with three portions (0.5 ml) of TFA, each wash being transferred to the glass funnel where it is mixed with the resin in order to ensure the complete extraction of any remaining peptide. Refrigerate the combined filtrates briefly (30–60 min) to complete peptide precipitation.

11. Collect the precipitate by centrifugation for 3–5 min at 2000–3000 g, and wash the solid thoroughly with small portions (5–10 ml) of diethyl ether or methyl t-butyl ether, each time the solid is collected by centrifugation. This process is repeated three to five times, or until all traces of scavengers are removed.

12. Dissolve the crude peptide in an appropriate aqueous solvent and lyophilize the solution.

---

[a] The cleaved and deprotected peptide will be in solution at this stage and will precipitate once it comes into contact with the ether layer, the spent resin being retained by the sintered glass.

ii. Use of trifluoromethane sulphonic acid

An alternative deprotection/cleavage strategy, based on the use of TFMSA, has been developed by Tam *et al.* (18) and, independently, by the Peptide

---

Protocol 9. 'Low–High' HF procedure

Equipment and reagents

- HF apparatus (*Figure 5*)
- Dimethyl sulphide (DMS)/*p*-cresol (26:4, v/v)
- Diethyl ether
- TFA

Method

1. Carry out the 'Low-HF' deprotection essentially according to steps **1–6** detailed in *Protocol 8*, with the following exceptions:

   - use a scavenger solution (30 ml) containing DMS/*p*-cresol into which is condensed HF (10 ml)[a]
   - allow the cleavage reaction to proceed at 0 °C for 2 h.

2. Evaporate the HF and scavenger solution at 0 °C.[b]

3. Repeat the cleavage using 'High-HF' conditions by following steps **1–6** detailed in *Protocol 8*, only using HF/*p*-cresol (10 ml, 9:1, v/v) for 30–60 min at −10 °C to 0 °C.[c]

4. Work the peptide up by following steps **8–12** in *Protocol 8*.

[a] If the peptide contains Trp(CHO), the formyl group can be removed under the 'Low-HF' conditions by replacing *p*-cresol with *p*-thiocresol or thiophenol.
[b] For peptides synthesized on MBHA or PAM resins, Tam *et al.* (17) recommend that the resin should be washed, on completion of this step, with DCM to remove sulphonium salts, dimethyl-sulphoxide (arising from the oxidation of DMS) and sulphur-containing reagents, prior to the 'High-HF' cleavage step.
[c] Peptides containing Arg(Tos) or Arg(NO$_2$) can be cleanly deprotected using this second step.

Research and Development group at Applied Biosystems Incorporated (19). Unlike the HF method, the use of TFMSA requires no specialized apparatus such as polytetrafluoroethylene reaction vessels, and the cleavage and side-chain deprotection reactions can be carried out in normal glassware. **Despite the improvement in the ease of use of TFMSA over HF, it should be appreciated that the former is an extremely corrosive liquid and must be handled with great caution. Consequently, all procedures must be carried out in an efficient fume cupboard; the use of rubber gloves—preferably gauntlets—and safety visor is to be considered mandatory.**

The procedures for the cleavage/deprotection of small quantities (∼ 100 mg) of resin-bound peptides by TFMSA are given in *Protocol 10*.

Unfortunately TFMSA does not provide a universal alternative to HF. *Table 2* lists those amino acid derivatives that are compatible with this cleavage method; however, there are quite a few side-chain protecting groups that cannot be cleaved cleanly. For example, the cyclohexyl ester derivatives of

---

Protocol 10. Small-scale TFMSA-mediated peptide cleavage and side-chain deprotection

Reagents

- Thioanisole/ethanedithiol (2:1, v/v)
- TFA
- TFMSA
- Diethyl ether

Method

1. Add 100 mg of resin-bound peptide to a 10 ml round-bottomed flask fitted with a microstirrer bar.

2. Add about 150 μl of thioanisole/ethanedithiol, fit a glass stopper, and stir for 10 min at room temperature.[a]

3. Add 1 ml of neat TFA and stir for 5–10 min.

4. Add TFMSA (100 μl) **dropwise** with vigorous stirring in order to avoid generation of heat. **Exercise great care in the handling of this acid.**

5. Allow the cleavage/deprotection to proceed for 15–30 min in the case of peptides anchored to a benzyl-type resin, otherwise 90–120 min is required for peptides linked *via* an amide bond to MBHA resins.

6. Follow steps **9–12** in *Protocol 8*.

[a] This operation and all subsequent steps **must** be carried out in an efficient fume cupboard.

---

aspartic and glutamic acids, as well as the MeBzl and Tos derivatives of cysteine and arginine, respectively, are refractory towards this acid and can only be removed cleanly by HF.

In a similar way to HF, TFMSA can be used in a standard 'one-pot' cleavage protocol or a 'Low–High' protocol depending on the amino acids contained within the sequence. The latter approach is particularly useful for the cleavage of peptides containing Met(O) and Trp(CHO). For example, under the 'low' conditions Met(O) can be reduced to Met by the inclusion of DMS in the cleavage solution, whereas addition of EDT will remove the formyl group from the indole nitrogen of tryptophan.

3. The Fmoc methodology

The 9-fluorenylmethyloxycarbonyl (Fmoc) group was introduced as a N-α-protecting group by Carpino and Han in the early 1970s (4). It can be removed by brief treatment with bases such as TEA and piperidine to generate the free base of the amino acid (see *Figure 6*). Indeed the exceptional lability of the Fmoc group to these bases ($t_{1/2}$ for cleavage with a 20% (v/v) solution of piperidine in DMF is less than 0.1 min), offered tremendous

Table 2. Side-chain protecting groups used in combination with N-α-Boc protection that are removed by TFMSA

Amino acid	Protecting group
Arginine	N-G-mesitylene-2-sulphonyl (Mts)
Lysine	N-ε-chlorobenzyloxycarbonyl (Cl-Z)
Histidine	π-4-toluenesulphonyl (Tos)
Cysteine	4-methoxybenzyl (Mob)
Tyrosine	O-2-bromobenzyloxycarbonyl (Br-Z)
Serine/threonine	O-chlorobenzyl (Cl-Bzl)
Aspartic acid/glutamic acid	O-chlorobenzyl (Cl-Bzl)
Met	sulphoxide [Met(O)[a]
Tryptophan	mesitylene-2-sulphonyl (Mts)

[a] Met(O) can be reduced quantitatively to methionine by including dimethyl sulphide in the TFMSA/TFA cleavage solution.

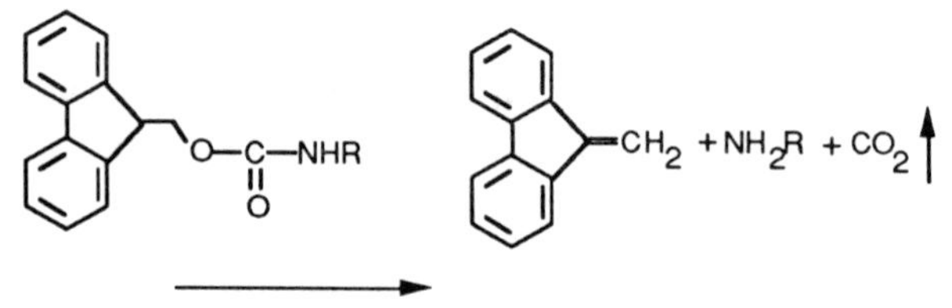

Tertiary base (piperidine, triethylamine, etc.)

Figure 6. Reaction scheme for the base-catalysed removal of the Fmoc group from amine functions. The base-catalysed cleavage of the Fmoc group from the amine NHR, generates the free amine NH_2R, dibenzofulvene, and CO_2.

potential for the development of an alternative chemical approach to the Merrifield solid-phase strategy that was not crucially dependent on repetitive acidolysis for the continued elaboration of the peptide chain. This potential was realized independently and virtually simultaneously by two groups in 1978. Thus Meienhofer and co-workers introduced a Fmoc-polystyrene-based protocol (5, 6), whereas Sheppard and colleagues developed a radically different approach that used beaded polydimethylacrylamide as a polar supporting matrix (20). Sheppard had argued as early as 1971 that the optimization of SPPS required the efficient solvation of the polymeric support and the growing peptide chain, and that these two processes could be enhanced by the use of an aprotic polar solvent such as DMF (21). Since it had already been established that formation of the peptide bond occurred some 4500-fold more readily in DMF than in chloroform or DCM (22), the use of DMF seemed to offer a convenient and convergent solution to these three processes.

The two generic Fmoc methodologies (i.e. Fmoc-polystyrene and Fmoc-polyamide) have been continuously refined and developed by numerous research groups over the subsequent 15 years (these have been recently reviewed by Fields and Noble (8)); however, the Fmoc-polyamide strategy has become synonymous with Sheppard's group at the Medical Research Council's Laboratory of Molecular Biology in Cambridge. This group has pioneered many of the developments regarding the application of Fmoc-protected amino acids to SPPS. Thus, in addition to introducing the use of polar solid-phase matrices such as polydimethylacrylamide suitable for use in the batch-wise protocols, they have also developed Kieselguhr-supported polydimethylacrylamide matrices for the synthesis of peptides under continuous-flow conditions, where the support and tethered peptide are always solvated (23). They have also been at the leading edge in the development of novel linkers for the mild acidolytic cleavage of the final peptide from the solid support (24, 25). Some of the practical aspects of these developments and those relating to the Fmoc-polystyrene methodology are discussed in the following sections.

3.1 General strategy

The lability of the Fmoc group to bases such piperidine provides an ideal basis for the development of a completely orthogonal protecting strategy ideally suited to SPPS. The avoidance of repetitive acidolysis for the generation of the free amino functions necessary for chain elongation (a central feature of the Merrifield Boc strategy), means that the side-chain protecting groups can be chosen so as to be removed by mild acid treatment upon completion of the synthesis and under conditions which can also effect cleavage of the peptide from the solid support. It can be immediately appreciated that the single most attractive feature of the Fmoc strategy is the avoidance of the use of HF (used to achieve side-chain deprotection and peptide cleavage in the Merrifield Boc strategy) and, consequently, the provision of the special handling facilities that the use of this acid requires.

The general scheme for Fmoc SPPS is shown in *Figure 7* and holds true for both the Fmoc-polystyrene and Fmoc-polyamide approaches. In essence the method can be reduced to the following steps.

(a) The initial N-α-Fmoc-protected amino acid is coupled to the solid support.

(b) The resin-bound Fmoc-amino acid is then deprotected by piperidine.

(c) The next Fmoc-amino acid is then coupled.

(d) The second and third steps are then repeated as necessary until the desired sequence has been constructed.

(e) The side-chain protecting groups and linkage between the solid support and peptide are then cleaved concomitantly by TFA to release the fully deprotected product.

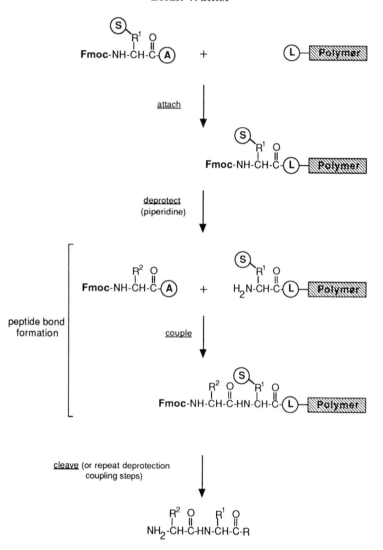

Figure 7. Flow diagram of the synthetic scheme for the solid-phase synthesis of peptides using the Fmoc strategy. After linking the first *N*-α-Fmoc-protected amino acid, via the carboxyl activated derivative **A**, to the resin-bound linker **L**, the peptide is elaborated by a repeated cycle of deprotection and acylation (coupling) steps. Amino acids bearing nucleophilic side-chain functional groups are incorporated using the orthogonal protecting group, **S**. Depending on the method used to cleave the peptide from the resin, the product can be obtained as a C-terminal acid (R = OH), ester (R = OMe, OEt, etc.), hydrazide (R = NHNH₂) or amide (R = NHMe, NHEt, etc.).

In order to carry out these steps, one can either adopt a batch-wise or continuous-flow strategy. If the former approach is used, then a simple bubbling device such as that utilized for Boc SPPS, and which is illustrated in *Figure 2*, can be employed for all the necessary synthetic manipulations. Alternatively, numerous types of apparatus are currently available for the continuous-flow approach for peptide construction. These range from simple manual through semiautomated to fully automated machines (see (26) for a comprehensive discussion of those instruments that are currently available). The author's lab currently synthesizes peptides on a semiautomated continuous-flow instrument, controlled by a microcomputer, similar to that designed by Dryland and Sheppard (27). Using such a set-up, one has to add the activated Fmoc-amino acid manually to the functionalized resin contained within the reaction column; however, the remainder of the steps such as Fmoc group removal, washing the resin with DMF after the deprotection and coupling steps, and recirculation of the activated Fmoc-amino acid derivatives during the coupling cycles, are performed automatically. The computer is programmed to interrupt the synthesis at the end of each deprotection and coupling step, in order for resin samples to be removed and analysed using the Kaiser ninhydrin and trinitrobenzene sulphonic acid (TNBSA) tests for free amino groups (see *Protocols 14* and *15*, respectively) to assess the completeness of each of these steps.

3.2 Chemical procedures

3.2.1 Choice of functionalized resins

Depending on the nature of the linkage, peptides can be obtained with a C-terminal amide ($-CONH_2$) or acid ($-CO_2H$) grouping, when cleaved with TFA, or as substituted amides ($-CONHR$) or hydrazides ($-CONHNH_2$) when treated with primary amines and hydrazine, respectively. *Figures 8* and *9* show examples of the more commonly used linkers that are employed in the preparation of peptide amides and peptide acids, respectively (see (8) for a more comprehensive listing).

The 4-hydroxymethylbenzoic acid linker (see *Figure 8*) is unique amongst those shown in these two figures, since it alone cannot be cleaved with TFA to release the final peptide. However, the ester bond between this linker and the peptide is readily cleaved by amine nucleophiles, and peptides can be obtained as their C-terminal primary amides by treatment with methanolic ammonia, or, by using a solution of a primary amine in methanol, as substituted amides (26).

The 4-hydroxymethylphenoxyacetic acid (HMPA) and 4-hydroxymethyl-3-methoxyphenoxyacetic acid (HMMPA) linkers shown in *Figure 9*, were introduced by Sheppard's group for the preparation of peptide acids, using Fmoc-polyamide batch-wise protocols (24, 25). The peptide–HMPA ester linkage can be readily cleaved with a solution of TFA in DCM (1:1, v/v, 30–60 min),

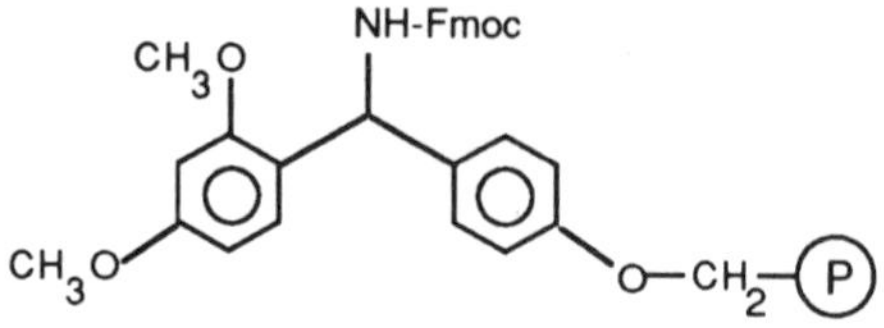

4-(2',4'-Dimethoxyphenyl-Fmoc-aminomethyl)-phenoxy-

(Rink amide)

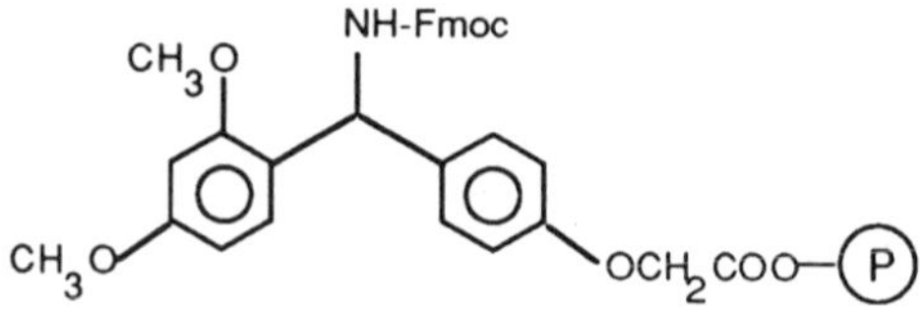

4-(2',4'-Dimethoxyphenyl-Fmoc-amniomethyl)-phenoxyacetic acid

(Sasrin-amide)

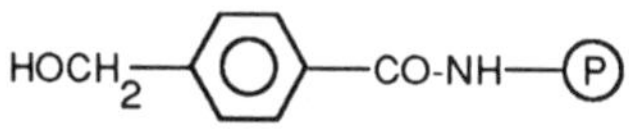

4-hydroxymethylbenzoic acid

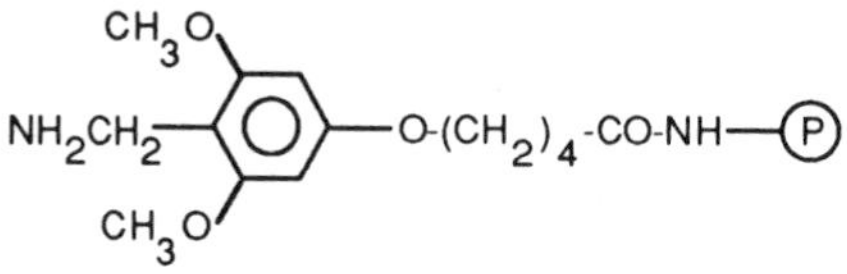

4-aminomethyl-3,5-dimethoxyphenoxybutyric acid (PAL)

Figure 8. Structures of the most commonly used linkers employed in the synthesis of peptide amides utilizing Fmoc SPPS. **P** represents the polymeric matrix which could be, for example, polystyrene- or polyamide-based.

conditions that will also result in the concomitant removal of any *t*-butyl-based side-chain protecting groups that may have been employed in the incorporation of trifunctional amino acids into the desired sequence (see Section 3.4.1). The HMPA linker most closely resembles the Wang linker (also shown in *Figure 9*), in terms of acid lability (28).

The incorporation of the methoxy substituent into the aromatic ring of HMPA substantially enhances the acid lability of the peptide-linker ester

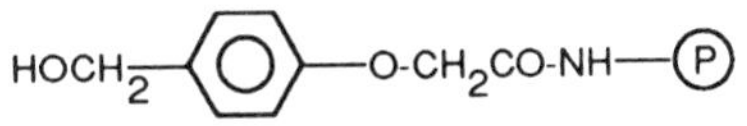

HOCH$_2$—⟨O⟩—O-CH$_2$CO-NH—Ⓟ   **Cleaved by TFA/DCM (50%, v/v)**

4-hydroxymethylphenoxyacetic acid (HMPA)

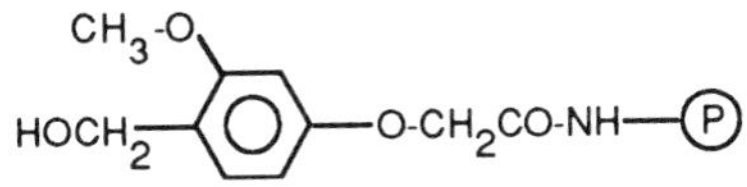

CH$_3$-O
HOCH$_2$—⟨O⟩—O-CH$_2$CO-NH—Ⓟ   **Cleaved by TFA/DCM (1%, v/v)**

4-hydroxymethyl-3-methoxyphenoxyacetic acid (HMMPA)

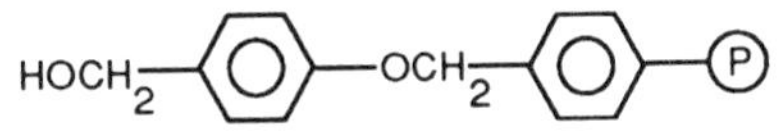

HOCH$_2$—⟨O⟩—OCH$_2$—⟨O⟩—Ⓟ   **Cleaved by TFA/DCM (50%, v/v)**

p- alkoxybenzyl alcohol polystyrene resin
(Wang resin)

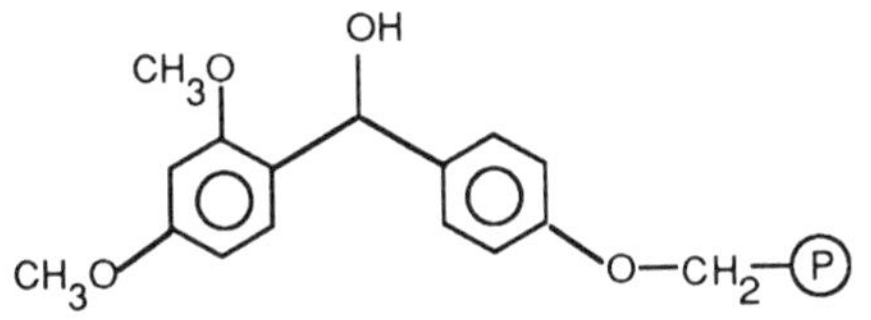

CH$_3$O, CH$_3$O — OH — O—CH$_2$—Ⓟ   **Cleaved by TFA/DCM (0.1%, v/v)**

4-(2',4'-dimethoxyphenyl-hydroxymethyl)-phenoxy-
(Rink acid)

Figure 9. Structures of the most commonly used linkers employed in the synthesis of peptide acids utilizing Fmoc SPPS. **P** represents the polymeric matrix which could be, for example, polystyrene- or polyamide-based.

bond. This linker was specifically devised by Sheppard so as to permit the fragment condensation of peptide acids with fully protected side-chain functionalities (29).

The Rink acid linker also shown in *Figure 9* is even more acid labile as it can be cleaved with 0.1% (v/v) solutions of TFA in DCM. Indeed, such is the acid lability of this peptide-linker ester bond that the Fmoc-amino acids to be coupled to the resin are sufficiently acidic to cause cleavage of the growing peptide chain from the solid support. This problem can be overcome, however, by using about 0.5% (v/v) solution of DIPEA in DMF during the coupling cycle (30).

The Rink amide, Sasrin amide, and 4-aminomethyl-3,5-dimethoxyphenoxy-butyric acid (otherwise referred to as PAL, an acronym for peptide amide linker) linkers (see *Figure 8* for structures) were originally introduced for

Fmoc SPPS using polystyrene-based supports and following a batch-wise protocol for peptide chain construction (30–32). However, there is now considerable 'methodology cross-talk' and both these acid-labile linkers have been used to functionalize Kieselguhr-supported polydimethylacrylamide beads for use in continuous-flow methodologies. These supports are commercially available from Novabiochem. Similarly, the HMPA linker is used with great success in Fmoc-polystyrene batch-wise protocols, and HMPA-polystyrene resins are commercially available from Applied Biosystems.

As mentioned above, one of the most radical developments in SPPS introduced by Sheppard was the adoption of a continuous-flow approach. Amongst the advantages cited for this approach over the batch-wise protocols employed by Merrified were: the more rapid and efficient removal of excess reagents, the avoidance of damage to the beaded gels arising out of attrition during agitation or stirring, which may be particularly prevalent in the synthesis of long peptides (>30 residues), and, arguably most importantly, ease of monitoring the synthesis (23, 27, 33–35).

By utilizing Kieselguhr as a supporting lattice onto which polydimethyl-acrylamide is grafted, the beaded polymer is sufficiently rigidified that it can resist compressional forces caused by pumping a solution through it. However, Kieselguhr particles are notoriously irregular in shape, and this gives rise to non-Newtonian fluid flow through the body of the matrix. This in turn leads to cavitation resulting in the physical attrition of the support and the generation of high back pressures. Despite the shortcomings of the Kieselguhr polyamide composites, the continuous-flow approach has become particularly popular among peptide chemists in Europe, and this has led to the development of improved support matrices for this particular approach. For example, Novabiochem, in conjunction with the National Starch and Chemical Corporation, have developed polyamide–polystyrene composites that possess greater mechanical strength than the Kieselguhr-supported polyamide resins. In these composites, a reticular polystyrene support (trade name Polyhipe) (36) is chemically bound to the polydimethylacrylamide gel thereby generating a macroporous resin that is rigid and non-compressible. These matrices, which are sold under the trade name NovaSyn-P Resins, have excellent flow characteristics and can be utilized for batch as well as continuous-flow systems, if desired. They are supplied with linkers suitable for the synthesis of peptide amides and peptide acids.

A similar approach has been taken by Bayer and Rapp in the development of polyethylene glycol (PEG) grafted polystyrene composites (37, 38). These supports represent a truly exciting development since the long PEG chains are so distant from the core of the supporting polystyrene matrix that the coupling of activated amino acid to the growing peptide chain can be considered as being effectively a solution-phase reaction. This means that the rates for the coupling reactions are almost equal to those in solution and are much more rapid than those observed for the acylation of peptides bound to 'standard' beaded polyamide or polystyrene supports. Additionally these supports can be purchased with linkers that give rise to peptide amides or acids, or in the form where the link between the PEG chain and the polystyrene can be cleaved by TFA thus releasing peptide–PEG conjugates that can be utilized directly in immunization protocols (38).

The protocols for the utilization of these various linkers and matrices for peptide assembly are discussed in the following sections.

3.2.2 Attachment of the first residue to functionalized resins

The synthetic protocols for the synthesis of the linkers used in Fmoc-polyamide SPPS and their subsequent utilization in resin functionalization have been described in great detail by Atherton and Sheppard (26), and will not be presented here. In any case, in common with all of the functionalized resins employed in this form of SPPS, the range of polyamide-derived resins shown in *Figure 8* is commercially available from various sources such as Applied Biosystems, Bachem, Novabiochem, MilliGen, and Peninsula Laboratories.

The methods employed for the attachment of the initial Fmoc-amino acid to the solid supports based on the polystyrene and polyamide beaded gel matrices are essentially identical to those discussed previously, in Section 2.2.1, for the Boc-protected amino acids. Consequently, this section will deal only with those loading protocols that differ substantially from those employed for the Boc-amino acids.

i. Resins containing amino-functionalized linkers

The Rink amide, Sasrin amide and PAL linker resins (see *Figure 8* for structures) are supplied as their Fmoc-protected derivatives; obviously the amino group of these linkers must be unmasked before the first Fmoc-amino acid can be attached to the solid supports. *Protocol 11* describes the procedures used for the removal of the Fmoc group from these linkers and also addresses the utilization of Fmoc-amino acid-pentafluorophenyl (Opfp) esters for the attachment of the first residue using a batch-wise process. *Protocol 12* should be utilized for the alternative continuous-flow approach.

Protocol 11. Attachment of first Fmoc-amino acid to amino-functionalized resins using Opfp active esters (batch process)

Equipment and reagents

- Manual bubbler (*Figure 2*)
- Amino-functionalized resin (available from a variety of suppliers such as Novabiochem, MilliGen and Bachem)
- Fmoc-amino acid-Opfp ester (available from Novabiochem and MilliGen)
- Piperidine
- DMF
- Acetic anhydride

Protocol 11. *Continued*

Methods

1. Place 0.5 g of resin (typical loading about 0.5 mmol/g) in the manual bubbler system shown in *Figure 2*.

2. Wash the resin with DMF (5 × 15 ml, 3 min per wash).

3. Wash the resin with a solution (15 ml) of piperidine in DMF (20% v/v, 1 min wash).

4. Wash the resin with a solution (15 ml) of piperidine in DMF (20% v/v, 15 min wash).

5. Wash the resin with DMF (5 × 15 ml, 3 min per wash).

6. Dissolve the Fmoc-amino acid-Opfp ester (0.75–1.0 mmol, representing a three- to four-fold excess over the resin-bound NH$_2$ groups) in DMF (~15 ml) and add this to the resin.[a, b]

7. Gently agitate the suspension for 45–60 min at room temperature.

8. Repeat step **2**.

9. If the Kaiser ninhydrin test (see *Protocol 14*) indicates that the reaction has not gone to completion, repeat steps **6** and **7** using a fresh solution of the Opfp ester.[c]

10. Should there be any remaining free amino groups after the repeat coupling, cap these by following step **8** in *Protocol 4*.

11. Repeat step **2**.

[a] Except for the arginine derivative which is unstable. The coupling of this residue can be achieved *via* its symmetrical anhydride derivative, prepared according to *Protocol 3*.

[b] The attachment of the first residue can be achieved equally easily by *in situ* activation methods employing HBTU or BOP chemistries (see *Protocol 13*), or by using pre-formed symmetrical anhydrides.

[c] Because of the base lability of the Fmoc group, it is generally better to limit reactions involving the coupling of Fmoc-amino acids to amine functions to a maximum of 60 min. The coupling reaction can then be repeated using freshly prepared reagents. This avoids possible inopportune removal of the Fmoc group.

Protocol 12. Attachment of first Fmoc-amino acid to amino-functionalized resins using Opfp active esters (continuous-flow process)

Equipment and reagents

See *Protocol 11*

Method

1. Load solvated resin (1 g/5 ml DMF) into the reaction column of an instrument of the type shown in *Figure 14*.

2. Wash the resin with DMF (flow rate 2.5 ml/min, 5 min).

3. Deprotect the resin with 20% (v/v) piperidine in DMF (flow rate 2.5 ml/min, 10 min).

4. Wash the resin with DMF (flow rate 2.5 ml/min, 15 min).

5. Dissolve the activated Fmoc-amino acid (three-fold excess over resin loading) in DMF (5 ml) and admit to the reaction column.

6. React resin-bound amino functions with activated Fmoc-amino acid derivative (recirculate for 60 min).

7. Remove a resin sample and assay for free amino groups using Kaiser or TNBSA tests (see *Protocols 14* and *15*, respectively).

8. Cap any remaining amino groups with 10% (v/v) acetic anhydride in DMF (recirculate for 30 min).

9. Wash resin with DMF (flow rate 2.5 ml/min, 8 min).

ii. Resins containing hydroxyl-functionalized linkers

The attachment of the first Fmoc-amino acid to resins functionalized with the hydroxyl-containing linkers in *Figures 8* and *9*, is carried out in the presence of catalytic amounts of DMAP, just as in the case for the attachment of Boc-amino acids to PAM resins. Numerous studies have been carried out to determine the best method of forming the ester bond between the linker and the Fmoc-amino acid, with regard to optimizing the loading of the resin and minimizing the racemization of the amino acid (39, 40). The use of pre-formed symmetrical anhydrides with 0.1 equiv. of DMAP in DMF has been reported to give high loadings (typically > 90%), using a 60 min reaction time, and low (< 1.0%) racemization (27, 41). The author's lab has generally found this method to be of the most widespread utility, except for the coupling of arginine, which generally gives a loading efficiency of about 45–55%. The batch-wise procedure detailed in *Protocol 2* for the attachment of Boc-amino symmetrical anhydrides to PAM resins is employed for the loading of the corresponding Fmoc derivatives to hydroxyl-containing linkers attached to beaded polystyrene and polyamide resins.

In addition to the use of pre-formed symmetrical anhydrides, loading with Fmoc-amino acid-Opfp esters in the presence of 0.2 equiv. of DMAP in DMF has been reported to give reasonable esterification (typically 75%, 1 h reaction time) and low levels of racemization (< 1%) (42).

If desired, the continuous-flow approach outlined for the attachment of activated Fmoc-amino acid derivatives to amino-functionalized resins (*Protocol 12*) can also be employed for the derivatization of suitable hydroxy-functionalized resins, with the proviso that the catalytic amount of DMAP be added to the activated derivative just prior to it being admitted to the reaction column containing the resin.

Table 3. Side-chain protecting groups that are most frequently used in combination with *N*-α-Fmoc protection

Amino acid residue	Protecting group
Arginine	*N*-G-2,2,5,7,8-pentamethylchromane-6-sulphonyl (Pmc)
	N-G-4-methoxy-2,3,6-trimethylbenzenesulphonyl (Mtr)
Lysine	*N*-ε-*t*-butyloxycarbonyl (Boc)
Histidine	*N*-τ-triphenylmethane (Trt)
	N-π-butyloxymethyl (Bum)
Cysteine	*S*-acetamidomethyl (Acm)
	S-triphenylmethane (Trt)
Tyrosine	*O*-*t*-butyl ether (*O*-tBu)
Serine/threonine	*O*-*t*-butyl ether (*O*-tBu)
Aspartic acid/glutamic acid	*O*-*t*-butyl ester (*O*-tBu)

3.3 Protection of side-chain functional groups

Table 3 lists some of the protecting groups for those nucleophilic side-chain functional groups of lysine, arginine, histidine, serine, threonine, cysteine, tyrosine, aspartic acid and glutamic acid, that are most commonly used in the Fmoc approach. The structures of some of these protecting groups are given in *Figure 10*. A brief discussion of the various protecting groups used for each amino acid is presented below.

i. Arginine

The most common ploy used to protect the guanidino function of arginine in the Fmoc strategy relies on the reduction of the nucleophilicity of this functional group by masking the ω-nitrogen by aryl (arene) sulphonyls such as 2,2,5,7,8-pentamethylchromane-6-sulphonyl-(Pmc) and 4-methoxy-2,3,6-trimethylbenzenesulphonyl-(Mtr) (see *Figure 10* for structures). Unfortunately, these protecting groups do not completely suppress the nucleophilicity of this side-chain, and activated derivatives of Fmoc.Arg(Pmc).OH and Fmoc.Arg(Mtr).OH undergo the same type of *intra*molecular cyclization reactions as Boc.Arg(Tos).OH and Boc.Arg(NO$_2$).OH$_2$, to form unreactive lactam derivatives. However, since a three- to four-fold excess of the activated derivatives is normally employed in the solid-phase approach, there is usually a surfeit of the desired activated derivative to ensure complete acylation of the resin-bound amine groups. Although the Mtr group can be removed cleanly, using TFA/thioanisole (95/5, v/v) for 30 min at room temperature, from peptides containing only one arginine residue (43), sequences containing multiple arginines require several hours exposure to the above cleavage mixture and, even then, all may not be removed cleanly (44, 45). The Pmc group was introduced by Ramage to circumvent this problem and it works very well for multiple arginine-containing peptides such as ubiquitin (46).

56

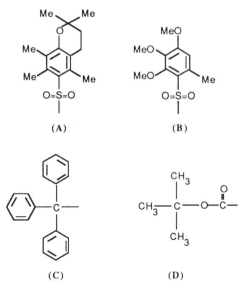

Figure 10. Structures of some common side-chain protecting groups utilized in Fmoc solid-phase peptide synthesis. A and B are 2,2,5,7,8-pentamethylchromane-6-sulphonyl- (Pmc) and 4-methoxy-2,3,6-trimethylbenzenesulphonyl- (Mtr) protecting groups respectively; these are used to mask the guanidino functional group of arginine. C and D are the triphenylmethane (Trt) and *t*-butyloxycarbonyl (Boc) protecting groups respectively, and are used to protect a variety of nucleophilic side chains of amino acids such as lysine (Boc), histidine (Trt and Boc), glutamine, asparagine (Trt), and cysteine (Trt).

ii. Lysine
The nucleophilic and basic ε-amino function of lysine is most commonly protected with the Boc group in Fmoc SPPS, although the polar trifluoro-acetyl group (Tfa) has also been utilized with great success, for example, in the synthesis of cytochrome *c* fragments (47). The Boc group can be removed upon completion of the synthesis with TFA, whereas the Tfa group, which is stable to both acid and mild bases, can be removed by strong alkaline hydrolysis (usually ammonium hydroxide). *N*-α-Fmoc.Lys(Fmoc).OH can be utilized in those circumstances where one wants to deliberately induce side-chain branching, by simultaneously deprotecting both the α- and ε-amino functions so that they can participate in subsequent coupling reactions. This ploy has been adopted in the so-called multiple antigen peptide (MAP) strategy, developed by Tam (48, 49) (see Chapter 4).

iii. Aspartic acid and glutamic acid
Aspartic acid and glutamic acid are most commonly incorporated as their β- and γ-*t*-butyl ester derivatives, respectively. The *t*-butyl esters, which are extremely resistant to nucleophiles, can be removed cleanly by TFA on completion of the synthesis.

iv. Asparagine and glutamine

Both Fmoc.Asn.OH and Fmoc.Gln.OH are almost completely insoluble in DCM and are only poorly soluble in DMF and *N*-methyl pyrrolidone. Additionally, the side-chain amide functions of both amino acids can undergo a dehydration reaction leading to nitrile formation, when their α-carboxyl groups are activated prior to coupling (see *Figure 11*) (50). Both of these problems can be circumvented by protecting the amide functionalities. Examples of amide protecting groups that are stable to bases such as piperidine but that can be removed by TFA are 2,4,6-trimethoxybenzyl (Tmob), 4,4′-dimethylbenzhydryl (Mbh) and triphenylmethane (Trt). The Fmoc-protected asparagine derivatives Fmoc.Asn(Tmob).OH (51), Fmoc.Asn(Mbh).OH (52) and Fmoc.Asn(Trt).OH (53) (and the corresponding glutamine analogues), are commercially available (Novabiochem, for example, sell the complete range), exhibit greatly improved solubility, and do not undergo the dehydration reaction discussed above.

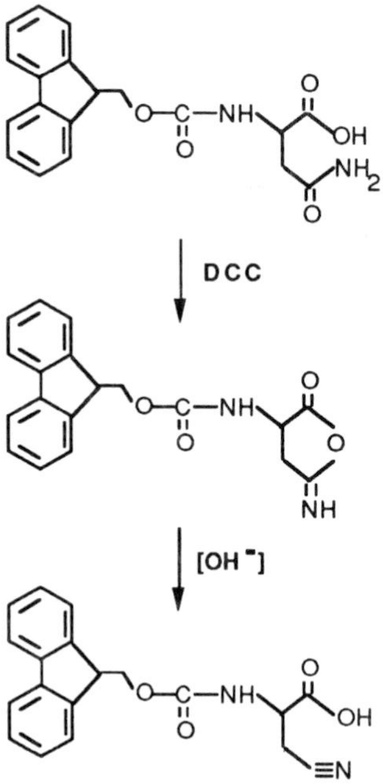

Figure 11. Reaction scheme for the dehydration of the β-amido group of asparagine that can occur during the activation of the α-carboxyl group. The generation of the nitrile derivative is thought to involve the formation of isoamide intermediate arising out of the intramolecular attack of the oxygen of the β-amide on the activated α-carbonyl grouping. Hydrolysis of this intermediate then generates the nitrile derivative as shown.

v. Serine, threonine, and tyrosine

Serine, threonine, and tyrosine are most commonly incorporated into peptide sequences as their *t*-butyl ether derivatives. The *t*-butyl group completely suppresses the nucleophilicity of the hydroxyl functions of these amino acids and can be removed by treatment with TFA. For peptides not containing histidine, it is possible to incorporate these amino acids without protecting their side-chain OH functions.

vi. Histidine

The most commonly used groups for the masking of the imidazole side-chain of histidine in Fmoc SPPS are shown in *Figure 12*. With the exception of the benzyloxymethyl group (Bom), which requires hydrogenolysis (H$_2$/Pd) or HF to effect its removal (54), these protecting groups can be removed cleanly and easily by TFA. Both the Bom group and the butyloxymethyl group (Bum) are incorporated into the imidazole ring at the π position, virtually eliminating the otherwise facile racemization of histidine residues during coupling (55). Similarly, the incorporation of the Boc group into the imidazole ring at the τ position also virtually eliminates racemization, whereas incorporation of the Trt group at this position, although reducing the basicity of the side-chain, does not completely eliminate racemization (8).

vii. Cysteine

The highly nucleophilic thiol side-chain of cysteine must be protected throughout the course of the synthesis. Failure to do so will result in the

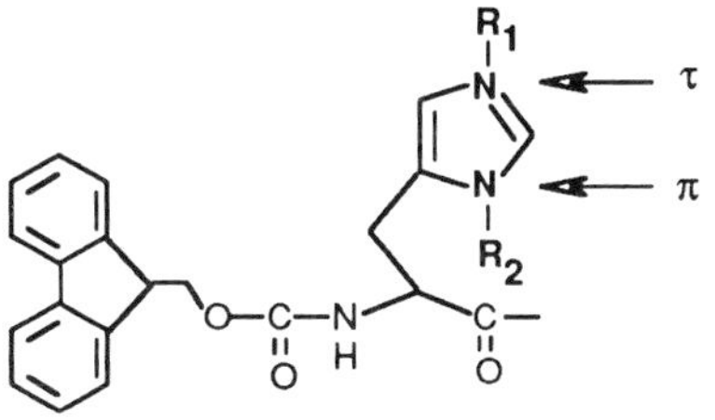

	N$^\tau$	N$^\pi$
	R$_1$ = Boc	R$_2$ = H
	R$_1$ = Trt	R$_2$ = H
	R$_1$ = H	R$_2$ = Bum
	R$_1$ = H	R$_2$ = Bom

Figure 12. The most commonly used combinations of protecting groups for histidine in Fmoc SPPS.

formation of mixed intra- and interchain disulphides and highly reactive thiol ester derivatives (formed by reaction between the side-chain thiol and activated Fmoc-amino acids) which could lead to the unpredictable and un-controlled incorporation of amino acids into the peptide sequence. The groups that have been most commonly used to protect the thiol function of cysteine in Fmoc SPPS are: acetamidomethyl (Acm), Trt, *t*-butyl (tBu) and *t*-butylsulphenyl (StBu). Both Acm and tBu are stable to TFA, whereas Trt and StBu can be cleaved by this acid (8). Acm is a particularly useful protecting group since it can be cleaved with mercuric salts to yield cysteine-containing peptides, or, in the case where there are multiple cysteine residues, by I_2 to yield cystine-containing peptides. Methods for forming specified disulphide bridges are discussed in Section 4.

3.4 Activation of amino acids and coupling procedures

In common with their Boc-protected counterparts, Fmoc-amino acids can be coupled via their symmetrical anhydride and active ester derivatives formed, for example, from pentafluorophenol and 1-hydroxybenzotriazole (HOBt). The preparation of symmetrical anhydride derivatives of *N*-α-Fmoc-amino acids can be achieved in an exactly analogous manner to their Boc-protected counterparts—essentially according to *Protocol 3*. The Fmoc derivatives of glutamine, asparagine, and tryptophan are particularly insoluble in DCM and the addition of 0.5–1.0 ml of DMF/1 mmol of each of these amino acids is normally required to aid solution. Alternatively, Fmoc.Asn/Gln(Tmob).OH (51), Fmoc.Asn/Gln(Mbh).OH (52) and Fmoc.Asn/Gln(Trt).OH (53) may be used to prepare the symmetrical anhydride derivatives of the first two amino acids, since, as mentioned earlier, they exhibit greatly enhanced solubility in DCM and are not prone to dehydration upon activation with carbodiimides.

Very recent attention has focused on the application of phosphonium and uronium activating reagents such as BOP (14) and HBTU (15) to Fmoc SPPS (56–58). These types of reagents, the structures of which are shown in *Figure 13*, are probably amongst the most active coupling reagents so far developed for peptide synthesis, and are now commercially available from many sources, but at a price!

In view of the possible carcinogenic potential of hexamethylphosphonamide, a by-product of the reaction of BOP with *N*-protected amino acids, the closely related derivative benzotriazole-1-yl oxy-tris-pyrrolidino-phosphonium hexa-fluorophosphate (PyBOP) was developed (59). This reagent is equipotent with BOP and is now the preferred phosphonium derivative. The protocols employed for the activation of Fmoc-amino acids with PyBOP and HBTU are identical, and are given in *Protocol 13*. This protocol also details the pro-cedure followed for the coupling of these activated derivatives to the resin-bound peptide/amino acid, using a batch process. Alternatively, the removal of the N-terminal Fmoc group and the subsequent washing and coupling steps

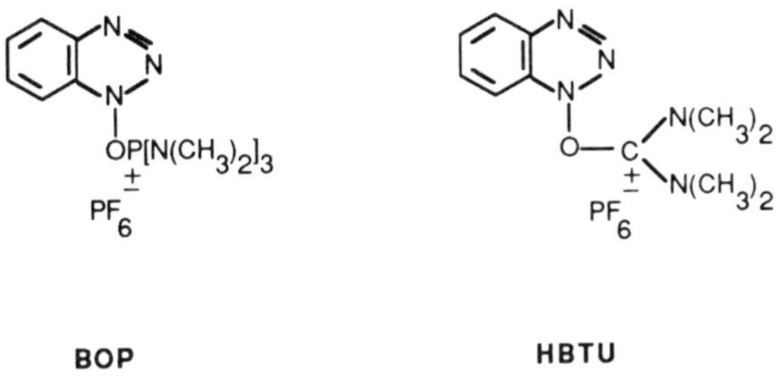

BOP   HBTU

Figure 13. Structures of benzotriazole-1-yl-oxy-tris-(dimethylamino)-phosphonium hexa-fluorophosphate (BOP) and 2-(1H-benzotriazole-1-yl)-1,1,3,3-tetramethyluronium hexa-fluorophosphate (HBTU).

can be carried out using the continuous-flow set up shown in *Figure 14* and by following the methodology outlined in *Protocol 12* for the attachment of the first activated Fmoc-amino acid to amino-functionalized resins.

Protocol 13. HBTU/BOP activation and coupling procedures (batch process)

Equipment and reagents

- Manual bubbler (*Figure 2*)
- Fmoc-amino acids
- HBTU or BOP
- DIPEA
- DMF
- Piperidine

Method

1. Place 0.5 g of Fmoc-amino acid or *N*-α-Fmoc-peptide-derivatized resin (typical loading about 0.5 mmol/g) in the manual bubbler system.

2. Wash the resin with DMF (5 × 15 ml, 3 min per wash).

3. Wash the resin with a solution (15 ml) of piperidine in DMF (20% v/v, 1 min wash).

4. Wash the resin with a solution (15 ml) of piperidine in DMF (20% v/v, 15 min wash).

5. Wash the resin with DMF (10 × 15 ml, 3 min per wash).

6. Dissolve the Fmoc-amino acid (0.75 mmol, three-fold excess over resin loading) in DMF (10 ml) and to this solution add HBTU or BOP (0.75 mmol) and DIPEA (1.50 mmol).

7. After 5 min, add the activated amino acid derivative to the resin and gently agitate the suspension for 30 min at room temperature.

8. Drain the resin and repeat step **2**.

Protocol 13. *Continued*

9. If the Kaiser ninhydrin, TNBSA (see *Protocols 14* and *15* respectively), or (in the case where the residue that is being acylated is proline) isatine (*Protocol 16*) tests indicate that the coupling has not gone to completion, repeat steps **6** and **7**.

10. Should there be any remaining free amino groups after the repeat coupling, cap these by following step **8** in *Protocol 4*.

3.5 Monitoring the extent of the acylation reactions

It is somewhat paradoxical that the major attractiveness of the solid-phase approach, i.e. the rapid synthesis of peptides without recourse to product-intermediate purification, also potentially represents its single most serious stumbling block. In postponing all of the isolation, purification, and charac-terization steps that are the trademark of classical organic solution-phase synthesis until the very last, the success of the solid-phase approach is singu-larly dependent upon the ability to drive each of the coupling and deprotec-tion steps close to 100% completion. *Table 4* shows the theoretical yield that can be expected for the synthesis of a peptide containing 5, 25, and 50 residues; it makes for rather sobering reading. For example, if we were to synthesize a peptide containing 25 residues and were fortunate to achieve a repetitive yield for each coupling and deprotection step of 95% (a synthetic efficiency that most classical organic chemists would envy), the best overall yield of product we could hope to obtain would be 8%. Of course in practice we would then have to attempt to purify the desired peptide from the crude product mixture containing the various truncated peptides and deletion sequences arising out of the incomplete coupling and deprotection reactions; this may well turn out to be an impossible task. For this reason, it is impera-tive that each step is driven as close to 100% completion as is possible; even so with an efficiency of 99% for each coupling and deprotection step, the maximum theoretical yield for a peptide containing 50 residues is only 36%.

Table 4. Theoretical percentage overall yield for the synthesis of peptides containing 5, 25, and 50 residues, assuming the given yields for each coupling and deprotection step

Peptide length	Repetitive yield		
	99%	95%	90%
5-mer	90	60	35
25-mer	60	8	0.5
50-mer	36	0.6	0.003

One of the great strengths of the solid-phase approach is that it permits the use of large excesses of reagents in an attempt at driving the coupling and deprotection reactions to completion; however, for one reason or another, it is often impossible to achieve this. It is imperative, therefore, that some type of monitoring or assessment of the extent of the acylation reactions and, ideally, deprotection steps be carried out, in order that any necessary remedial action (such as repeating the coupling and/or deprotection steps, or capping any free amino functions) can be taken to improve the efficiency of these processes. The protocols for the most useful colormetric tests and on-line real-time monitoring methods are discussed below.

3.5.1 Colormetric tests

The Kaiser ninhydrin (60) and TNBSA (61) colormetric tests (see *Protocols 14* and *15* respectively) for determining the amount of resin-bound amino groups, are rapid, sensitive, and easily carried out procedures that are ideally suited for monitoring the extent of acylation (coupling) and deprotection reactions. When resin loadings of about 0.5 mequiv/g are used, both reagents are capable of detecting the presence of about 0.5 μmol of amino groups/g of resin; this, for example, would correspond to a coupling efficiency of 99%. *Table 5* gives a 'rule of thumb' guide to the relationship between the extent of colour formation, using the Kaiser ninhydrin reagent, and the extent of coupling for polystyrene resins functionalized with peptide at a loading of about 1.0 mequiv/g. With polystyrene resins using both Boc and Fmoc strategies, the majority of the amino acids give strong responses with the ninhydrin reagent, except proline which, being an imino acid, cannot undergo the necessary reaction required for the formation of the blue chromophore; however, the istatin test (*Protocol 16*) can be used in this instance (62). Additionally, it has been our experience that serine, threonine, aspartic acid, and asparagine give variable responses depending on their location in the peptide sequence. **For the synthesis of peptides on polyamide beaded gels or Kieselguhr-supported polyamide composites, the Kaiser test is quite hopeless and should not be relied upon.**

Table 5. Relationship between extent of colour formation using the Kaiser ninhydrin test and extent of coupling reaction

Ninhydrin colour		% reaction
Beads	Solution	
Dark blue	Dark blue	76
Dark blue	Moderately blue	84
Moderately blue	Slightly blue	94
Slightly blue	Trace of blue	99

Protocol 14. Kaiser ninhydrin test

Reagents

- Ninhydrin
- Phenol
- KCN
- Ethanol
- DMF

Method

1. Make up the reagents required for this test by preparing the following solutions:

   - ninhydrin (1 g) dissolved in ethanol (10 ml)
   - phenol (80 g) dissolved in ethanol (20 ml)
   - KCN (2 ml of a 0.001 M solution in water) made up to 100 ml with pyridine. **KCN is a severe poison and should be weighed out on a balance situated in an efficient fume cupboard. Rubber gloves should be worn whilst preparing the aqueous stock solution of this reagent.**

2. Transfer 5–10 mg of the resin sample to be tested and place it in a 2 ml Eppendorf tube.

3. Add DMF (1 ml) to the sample, shake the tube for a few seconds to wash the beads, spin the tube at a high speed setting on microfuge for 30 sec, and finally remove the supernatant. Repeat this step three more times.

4. As in step **2**, only using ethanol (1 ml) in place of DMF.

5. Transfer a few of the beads to a clean Eppendorf tube (1 ml) and add two to three drops of each of the reagents prepared in step **1**.

6. Place the tube in a heating block or oven maintained at 120 °C for 5 min.

7. A positive result is obtained when the solution gives an intense blue-purple colour and, depending on the number of free amino groups present, the resin particles may also be intensely coloured. A negative test yields a straw-yellow colour with no coloration of the resin beads (see *Table 5*).

Protocol 15. Trinitrobenzene sulphonic acid (TNBSA) test

Reagents

- DIPEA/DMF (10% v/v)
- TNBSA

Method

1. Make up the reagents required for this test by preparing the following solutions:
   - DIPEA in DMF; make up fresh solution daily
   - TNBSA (5 mg) dissolved in above solution (0.5 ml) and prepared immediately before the test is to be carried out
2. Place a few of the beads to be tested in the depressions of a white porcelain test plate and add three to five drops of TNBSA solution.
3. The presence of free amino groups will result in the beads becoming strongly orange-coloured, within a few seconds of the solution being added. If no free amino groups are present, the beads will remain colourless.

Protocol 16. Isatin test for proline

Reagents

- Ninhydrin reagents (see *Protocol 14*)
- Benzyl alcohol
- Boc.Phe.OH
- Isatin
- Acetone

Method

1. Make up the reagent for the isatin test by preparing the following solutions:
   - isatin (2 g) dissolved in benzyl alcohol (60 ml). Vigorous stirring will be required to achieve complete solution
   - add Boc.Phe.OH (2.5 g) to the above solution
2. Place a few beads of resin sample to be tested in a 2 ml Eppendorf tube and add two to three drops of the isatin solution followed by two to three drops of each of the ninhydrin reagents.
3. Place the tube in a heating block maintained at 100 °C for 5 min.
4. Cool the tube to room temperature, decant the deeply coloured supernatant, and wash the beads with acetone by following the procedure outlined in step **3** of *Protocol 14*.
5. A positive result is obtained if the beads are a blue/red colour, otherwise a negative test is indicated if they remain colourless.

3.5.2 On-line monitoring

Although each of the above colorimetric tests can provide invaluable information as to the progress of the synthesis, they use up resin and slow the course of the synthesis as each step is checked for completion before proceeding to the next. This prompted Sheppard to introduce an on-line monitoring method for determining the extent of acylation of resin-bound amino groups by active esters of *N*-α-Fmoc-protected amino acids derived from 3-hydroxy-4-oxo-3,4-dihydro-1,2,3-benzoxytriazine and *p*-nitrophenol (34). Acylation of amino groups by these ester derivatives generates acid species the conjugate bases of which are highly coloured. Thus, provided acylation is incomplete, this will give rise to a colour change on the resin as a result of ionization induced by unreacted amino groups of the growing peptide, thus permitting continuous monitoring of the extent of reaction, either observing the reaction column 'by eye', or by measuring its absorbance with a specialized spectrophotometer (34). Although this is a particularly attractive approach since it gives a direct indication of the extent of amino group acylation in 'real-time', it has one obvious and major drawback in that it can only be applied to those activated esters that generate chromophoric acids. This led N. V. McFerran and the author to investigate the potential of measuring changes in the conductance of the reaction solution during acylation reactions in continuous-flow Fmoc SPPS (63, 64), an approach that was also taken independently by Nielsen *et al.* (65). The rationale behind our approach is that all forms of acylation reaction using symmetrical anhydrides or activated esters or acid chlorides/fluorides, generate acid components upon formation of the amide bond. Generally speaking, in aprotic solvents used for peptide synthesis, these are not ionized; however, addition of DIPEA to these acids will cause ionization in the polar solvents advocated for maximal peptide and resin solvation. The resultant ion pairs give rise to readily detectable conductance signals which are directly proportional to the amount of acid generated, thus permitting direct monitoring of the acylation reactions.

Using a semiautomated instrument equipped with a conductivity cell, a simplified diagram of which is shown in *Figure 14*, the progress of each coupling and deprotection reaction is monitored conductiometrically (64). In

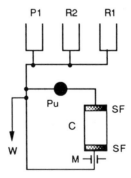

Figure 14. Simplified diagram of a continuous-flow reactor system. A description of the individual parts of this system is given in the text.

this diagram, **R1** is the DMF solvent reservoir, **R2** the piperidine/DMF reservoir, **P1** the port for the addition of activated Fmoc-amino acid derivatives (a 5 ml glass hypodermic syringe with a coarse sintered glass filter at the delivery end), **W** the exit to waste, **C** the reaction column (glass/Teflon, type 44–6111; Omnifit) with sintered frits (**SF**) at each end, and **Pu** the pump (LDC), set to a flow rate of 2.5 ml/min. **M** is a conductivity measuring cell. The control valves, marked by black circles are interconnected by Teflon tubing. Overall control is achieved by a suite of programs running on an Apple II microcomputer.

Some indication of the usefulness of this approach is given in *Figure 15*, which illustrates the changes in conductivity observed in coupling Fmoc.Ala. Opfp to Fmoc-Leu-HMP resin. The synthesis of this dipeptide follows the general scheme for Fmoc SPPS; in essence this involves Fmoc group removal followed by acylation of the free amino groups. It can be seen from *Figure 15* that both the initial deprotection and subsequent coupling phases (steps **P** and

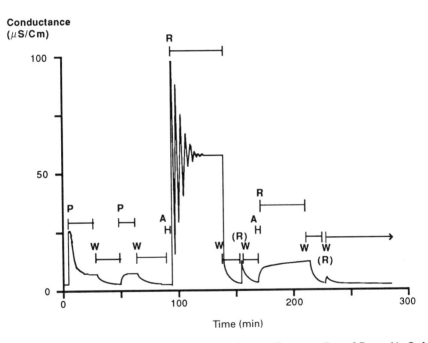

Figure 15. Complete conductivity profile for continuous-flow coupling of Fmoc.Ala.Opfp to Fmoc.Leu-HMPA resin. The processes shown are deprotection of the Fmoc.Leu-HMP resin followed by coupling of Fmoc.Ala.Opfp in a continuous-flow reactor system of the type shown in *Figure 14*. Each of the deprotection steps are indicated by the periods **P**, washing cycles by the periods **W**, and coupling steps by the periods **R**. The introduction of the activated ester to the reaction column is achieved during period **A**. The washing out of the dead volume between the two valves immediately upstream of the pump (see *Figure 14*) is represented by the period (**R**).

R, respectively) give rise to characteristic but clearly different signals; the different nature of the two signals reflects the difference in the reaction environments for the two processes. Removal of the Fmoc group being achieved by permitting bulk piperidine-containing solution to flow through the resin to waste, whilst coupling of the activated amino acid is carried out by dissolving it in a small volume (2–5 ml) and admitting it to the reaction column (step **A**), where it is recirculated in order to effect acylation of the amino groups (step **R**).

Deprotection thus gives rise to a single spike of conductance (the conductive species is probably a piperidine-carbamic acid adduct arising out of reaction of CO_2, generated from Fmoc group decomposition, with piperidine) with a very steep leading edge, indicating very fast liberation of the majority of the Fmoc groups; the falling edge is less steep, perhaps indicating slower deprotection of a minority of the resin-bound molecules. Following a washing cycle (step **W**) in which the conductance falls to its base-line value, reapplication of the piperidine deblocking solution to the resin (step **P**, 50–70 min in *Figure 15*) does not show the characteristic peak of an active deprotection; the monotonic increase simply reflecting the differences in conductance between neat DMF and a 20% (v/v) solution of piperidine in DMF.

In the acylation phase (step **R**, 90–120 min), diffusion throughout the total reactor volume of both the concentrated solution of Fmoc.Ala.Opfp and an initial narrow zone of pentafluorophenolate:DIPEA adduct, arising out of acylation of the resin-bound amino groups, give rise to the characteristic decaying spikes in the conductance signal. A second application of the active ester (step **R**, 170–220 min), gives an almost monotonic increase, with slight 'ripples' on the leading edge of the conductance signal due solely to the dilution, in the absence of any further acylation, of the concentrated reagent in the larger volume of the reaction column. The small conductivity spikes marked by (**R**) following the coupling cycles reflect washing out of the 'dead limb' of the apparatus between the two valves immediately upstream from the pump.

This simple example clearly demonstrates the utility of the conductance method not only for following the acylation step in the synthesis in 'real-time' but also for monitoring the removal of the Fmoc protecting group. This method has been successfully applied to the monitoring of the synthesis of more complex peptides and it has been particularly useful in demonstrating unequivocally that the removal of Fmoc groups has occurred, which the ninhydrin method has singularly failed to do so (unpublished results).

3.6 Cleavage of peptide from resin and side-chain deprotection

One of the most attractive features of the Fmoc SPPS approach is that the constructed peptide can be cleaved from the solid support under relatively mild acid conditions using TFA. Also, as outlined in Section 3.3, considerable attention has been paid to the development of side-chain protecting groups that can be removed under similar acid conditions, thus permitting, if desired, the concomitant deprotection and cleavage of the peptide from the solid support. However, although much milder than the methods employed to deprotect and cleave peptides synthesized by the Boc SPPS strategy (i.e. HF or TFMSA), the use of TFA is not without its drawbacks. For example, the TFA-catalysed cleavage of the Tmob, Trt, and Mbh groups from asparagine and glutamine, the Mtr and Pmc groups from arginine, the *t*-butyl-based groups for tyrosine, serine, threonine, and aspartic acid, and glutamic acid, and the Trt from cysteine, results in the generation of electrophiles that can alkylate residues such as tryptophan, methionine, cysteine, and tyrosine (45). In addition to this, each of the resins listed in *Figures 8* and *9*, in which the peptide-linker bond can be cleaved acidolytically, also have the potential for generating resin-bound carbocations. This can be particularly problematical if the C-terminal residue is tryptophan, since alkylation of this amino acid by such linker-derived electrophiles can result in the irreversible covalent attachment of the peptide to the matrix (26).

Fortunately, side reactions such as these can be greatly reduced by the inclusion of carbocation scavengers in the deprotection mix, the exact composition of which depends on the peptide sequence. *Protocols 17* to *19* give details of the procedures and scavengers used in the cleavage and deprotection of peptide containing a **C-terminal acid**, synthesized on any resin **apart** from Rink acid or Sasrin. In essence, the protocols are very similar to one another; however, each uses a different mix of scavengers in the deprotection/cleavage mix. For example, *Protocol 17* is used for peptides not containing carbocation-sensitive amino acids, whereas *Protocols 18* and *19* are recommended for sequences containing any one of tyrosine, methionine, or cysteine, and tryptophan and/or arginine, respectively.

Protocol 17. TFA cleavage and deprotection of peptides not containing problematic amino acids

Equipment and reagents

- Glassware for suction filtration to include: sintered glass funnel (No. 1 or 2 porosity), conical flask (with side arm) and round-bottomed flask (50–100 ml) These are all available from Quickifit
- TFA

- Methanol
- Diethyl ether
- DCM
- P_2O_5
- KOH

Method

1. Place the resin-bound peptide in a sintered glass funnel fitted to a conical flask with side arm. Apply gentle suction to flask by way of a water pump.

Protocol 17. *Continued*

2. Wash the resin with DCM (5 × 50 ml).

3. Wash the resin with methanol (5 × 50 ml).

4. Dry the resin under high vacuum, over P_2O_5, for at least 4 h (preferably overnight).

5. Place the washed and dried resin in a round-bottomed flask (50–100 ml, depending on scale) and add a 95% (v/v) solution of TFA in water (25 ml/ 0.2 mequiv of resin).[a] Allow the cleavage/deprotection reaction to proceed at room temperature for 2 h, with occasional stirring.

6. Filter the cleavage mixture, using a sintered glass funnel, and concentrate the filtrate, using a rotary evaporator, to 1–2 ml. Keep water bath temperature at less than 20 °C.

7. Split the TFA concentrate between two 10 ml glass centrifuge tubes and add diethyl ether (about 8 ml) to each. Refrigerate briefly (30–60 min) to complete peptide precipitation.[b]

8. Collect the precipitate by centrifugation for 3–5 min at 2000–3000 g, discard the supernatant, and wash the pellet thoroughly with a small portion (about 8 ml) of diethyl ether. This process is repeated three to five times, each time the solid is collected by centrifugation.

9. Dry the crude peptide over KOH pellets, under high vacuum, overnight; this is essential in order to remove any residual traces of TFA.

---

[a] **CAUTION**. For this, and each of the protocols listed below that involve the use of TFA, it must be appreciated that this acid is an extremely corrosive liquid, and must be handled with great caution. Consequently, all such procedures must be carried out in an efficient fume cupboard; the use of rubber gloves—preferably gauntlets—and safety glasses is to be considered mandatory.

[b] Very hydrophobic peptides frequently require longer refrigeration periods to ensure complete precipitation, in these instances it is better to err on the side of caution and leave the precipitate to form overnight.

Protocol 18. TFA cleavage and deprotection of peptides containing tyrosine, cysteine, or methionine

Equipment and reagents

- As in *Protocol 17*
- EDT

Method

1. Wash the resin thoroughly to remove any traces of DMF by following steps **1–4** of *Protocol 17*.

2. Place the washed and dried resin in a round-bottomed flask (50–100 ml, depending on scale) and add a solution (25 ml/g of resin) containing 95% TFA/5% EDT (v/v).[a,b] Allow the cleavage/deprotection reaction to proceed at room temperature for 2 h, with occasional stirring.

3. Work up peptide as described in steps **6–9** in *Protocol 17*.

[a] This step and the subsequent ether washings must be carried out in an efficient fume cupboard as EDT is a foul-smelling liquid.

[b] For peptides containing residues protected with the Trt group (*i.e.* asparagine, glutamine, cysteine, or histidine), the inclusion of 2–5% (v/v) triisopropylsilane is recommended for the efficient scavenging of trityl carbocations.

Protocol 19. TFA cleavage and deprotection of peptides containing tryptophan or arginine

Equipment and reagents

- As in *Protocol 17*
- Reagent K (82.5% TFA/5.0% H_2O/5.0% thioanisole/5.0% phenol/2.5% EDT, v/v/v/v/v)

Method

1. Wash the resin thoroughly to remove any traces of DMF by following steps **1–4** of *Protocol 17*.

2. Place the washed and dried resin in a round-bottomed flask (50–100 ml, depending on scale) and add Reagent K (25 ml/g of resin).[a] Allow the cleavage/deprotection reaction to proceed at room temperature for 2 h, with occasional stirring (66).[b]

3. Work up the peptide as described in steps **6–9** in *Protocol 17*.

[a] The addition of 2–5% triisopropylsilane is also recommended if any residues have been protected with the Trt group.

[b] For peptides containing more than one arginine residue, and particularly in those instances where this amino acid has been incorporated as the Mtr derivative, extended reaction times (4–6 h) will be required to achieve complete deprotection.

For the cleavage of peptides from Rink amide resin *Protocol 20* should be followed. In the original method, Rink recommended the use of 2% TFA in DCM (v/v) for the cleavage of peptide from the solid phase (30); however, the linker between peptide and resin is not as labile as was originally assumed and efficient cleavage can only be achieved using solutions containing more than 5% (v/v) TFA. Generally speaking, it is better to stick to solutions containing between 5 and 10% (v/v) TFA, as higher concentrations can lead to the fission of the linker from the support giving rise to soluble carbocations that can alkylate sensitive residues such as tryptophan.

Once the peptide has been cleaved, simple filtration is used to free it from spent resin and, depending on which amino acids are present, the protecting groups can be then removed by using one of the appropriate TFA/scavenger cocktails detailed in *Protocols 17–19*.

Protocol 20. Cleavage and deprotection of peptides from Rink amide resin

Equipment and reagents

• As in *Protocol 17*

Method

1. Wash and dry the resin as detailed in steps **1–4** of *Protocol 17*.

2. Slurry the washed and dried resin with 10% solution of TFA in DCM (10–15 ml/g resin) and transfer to a sintered glass funnel (porosity No. 3 or 4).

3. Allow the acid solution to slowly percolate through the resin and collect the filtrate in a round-bottomed flask (100–250 ml).

4. Wash the resin with 5% solution of TFA in DCM (20–30 ml), once again allowing the solution to percolate through the resin bed, and combine the washings and filtrate.

5. Concentrate the cleavage solution by rotary evaporation, ensuring that the water-bath temperature does not exceed 20 °C.

6. For the removal of the side-chain protecting groups, follow steps **5–9** in *Protocol 17* if the peptide contains no problematic residues, steps **2–3** in *Protocol 18* if it contains tyrosine, cysteine, or methionine, or steps **2–3** in *Protocol 19* if it contains tryptophan or arginine.

In the light of the potential problems associated with the lability of the linker in the original Rink amide resin, improved versions have been developed in which the bond between the linker and the polymer has been rendered completely stable to acidolytic fission (see *Figure 16*). If these are utilized for the synthesis of peptides containing C-terminal amides, the peptides can be cleaved from the solid support and deprotected in one step using one of *Protocols 17–19*, depending on the particular amino acid sequence.

The preparation of fully protected peptides containing C-terminal acids suitable for use in fragment condensation studies, can be achieved using solid supports containing Rink acid, Sasrin or 4-hydroxymethyl-3-methoxyphenoxyacetic acid (HMMPA) linkers. The ester bond formed between the first amino acid of the peptide and each of these linkers can be cleaved with very dilute solutions of TFA (typically 1% (v/v) in DCM) to release the

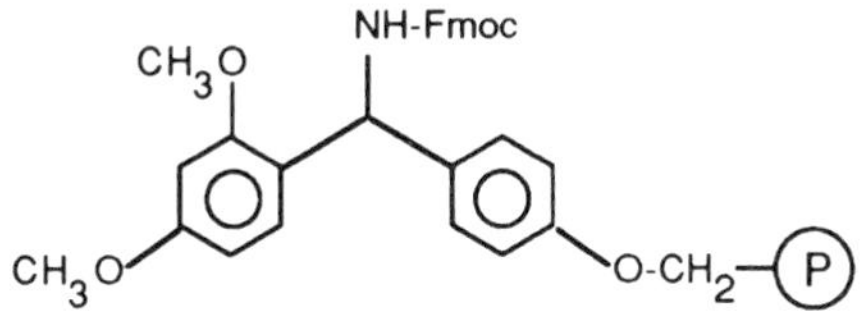

Rink-amide

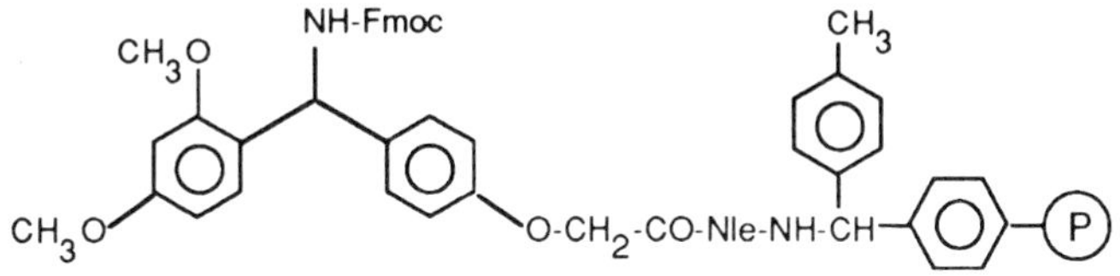

Rink-amide-MBHA

Figure 16. Structure of the improved Rink amide MBHA linker. Nle, norleucine.

globally protected peptide fragment into solution. *Protocol 21* outlines the procedure used for the cleavage of such peptides; it can be employed for each of the linkers listed above.

Protocol 21. Cleavage of fully protected peptides from resins possessing high sensitivity to acids (Rink acid, Sasrin, HMMPA, etc.)

Equipment and reagents

- As in *Protocol 17*
- Methanol/pyridine (1:250 v/v).

Method

1. Wash and dry the resin as detailed in steps **1–4** of *Protocol 17.*[a]

2. Slurry the washed and dried resin with 1% solution of TFA in DCM (10 ml/g resin) and transfer to a sintered glass funnel (porosity No. 3 or 4).

3. Allow the acid solution to slowly percolate through the resin and collect the filtrate in a round-bottomed flask (500 ml), containing a solution (250 ml) of methanol in pyridine.[b] Repeat this process a further three to four times,[c] using fresh batches of TFA solution, each time collecting the filtrate in the same pyridine solution.

73

Protocol 21. *Continued*

4. Remove the solvents *in vacuo* and treat the residue with a 10% aqueous solution (v/v) of acetic acid.

5. Collect the product by filtration and dry in high vacuum, over KOH pellets, overnight.

[a] Since only very dilute solutions of TFA are employed in this protocol, it is important that all traces of DMF be removed from the resin as it will severely retard the cleavage process. Therefore, it is imperative that these washing steps be carried out thoroughly.
[b] Alternatively, the TFA can be neutralized using a solution of DIPEA in DCM (use an equivalent amount of the base dissolved in DCM, about 150 ml).
[c] Depending on the buffering capacity of the particular peptide sequence, the majority of the product could be contained in the first or subsequent batches of acid filtrate; the extent of cleavage should therefore be monitored by thin-layer chromatography.

4. Methods of disulphide bridge formation

In many instances, the construction of the peptide chain on the solid support can turn out to be the least demanding step in the overall synthetic scheme; this can be particularly true in those instances where the sequence contains multiple cysteine residues. With such peptides, the preparation of the correctly folded final product can be crucially dependent on the formation of the appropriate disulphide bridging pattern and is often bedevilled by the generation of disulphide-linked *inter*chain polymers and incorrectly folded monomers containing undesired *intra*molecular cystine pairings. Added to this is the potential for chemical damage to the peptide arising out of oxidation of sensitive residues such as methionine and tryptophan by disulphide-forming reagents such as I_2 and $K_3Fe(CN)_6$ (67).

The first problem can largely be solved by carrying out the oxidation reaction using dilute solutions of peptide (typically 0.5–1.0 mM), thus favouring intramolecular over intermolecular disulphide formation. However, preventing the formation of undesired intramolecular cystine pairings can present a much more intractable problem, since each individual peptide will possess inherently different tendencies for disulphide scrambling, and the resulting incorrectly folded peptides can often aggregate and precipitate from solution.

Recent studies by Otaka *et al.* (68) and Tam *et al.* (69) on the utilization of dimethyl sulphoxide (DMSO) for disulphide bridge formation hold great promise, not only for the amelioration of the first two problems but also for the formation of disulphide bridges without oxidative damage to sensitive amino acids. Thus, Tam has demonstrated that the use of 20–50% (v/v) solutions of DMSO in a variety of buffer systems greatly promotes disulphide bond formation in comparison with other methods such as aerial oxidation. Additionally, unlike I_2 and $K_3Fe(CN)_6$ which are known to cause the conversion of methionine into methionine sulphoxide, DMSO-mediated disulphide bridge formation was found to have no affect on this amino acid and a further advantage was that DMSO was also found to greatly reduce and, in some instances, suppress completely, the aggregation and precipitation of peptides that occurred using other oxidative procedures.

The methodology employed by Otaka *et al.* (68) is, perhaps, of even greater potential since they have demonstrated that it is possible to effect the concomitant cleavage of cysteine side chain protecting groups such as S-Trt and S-Mab, and formation of disulphide bridges using solutions of DMSO in TFA (10%, v/v).

5. Purification of peptides

By far the most favoured method for purifying crude peptide mixtures, obtained after cleavage from the solid support and removal of the side-chain protecting groups, is reverse-phase HPLC. The use of C_{18} reversed phase columns has gained widespread popularity in this regard; however, for very hydrophobic peptides the recovery of material from C_8 columns is generally superior. Resolution of the components is usually achieved using either isocratic or gradient systems composed of two solvents, the most popular combination being 0.05% (v/v) TFA in water and 0.05% (v/v) TFA in acetonitrile. Elution of the products from the column is monitored by measuring the absorbance of the eluent at 210–220 nm, or, if the peptide contains aromatic amino acids, at 240–280 nm. Although these acidic solvent mixtures will readily dissolve hydrophilic and basic peptides, it may be necessary to dissolve very hydrophobic peptides in small volumes of DMF or propan-1-ol prior to loading on to the HPLC column. Similarly, acidic peptides can be dissolved by the addition of dilute ammonia.

In view of the great variety of HPLC systems currently available, and given the diversity of peptide sequences that may be synthesized, no methodological protocols will be presented for the purification of peptides using reverse-phase HPLC; however, the reader is referred to (70) and (71) for a general discussion of the method.

For peptides containing a preponderance of either acidic or basic residues, ion-exchange chromatography offers an excellent alternative to reverse-phase HPLC. Ion-exchange chromatography generally gives a high recovery of peptides and can be easily scaled up so as to process gram quantities of material. There is a great variety of ion-exchange materials available both for the purification of peptides containing a surfeit of basic amino acid residues (cation-exchange chromatography), for example, CM Sephadex (Pharmacia) and Whatman CM52 and those containing a surfeit of acidic amino acid residues (anion-exchange chromatography), for example, DEAE Sephadex and Whatman DE52. Since the conditions used to carry out the ion-exchange chromatographic separation will change from peptide to peptide and will also vary with the ion-exchange medium employed, no protocols will be presented here for the use of this particular technique. Instead, the reader is referred to (71) which gives detailed instructions on the preparation and use of Whatman cationic and anionic exchange materials for the purification of synthetic peptides.

5.1 Characterization of purified peptide

Having purified the synthetic product, its identity has to be established. By far the most commonly employed method for this purpose is amino acid analysis. This will establish whether or not the product contains the desired amino acids and that they are present in the correct ratio to one another. Alternatively, provided that the product is homogeneous, the determination of its molecular weight by fast atom bombardment mass spectrometry (FAB-MS), or electrospray mass spectrometry (ES-MS), is also considered as proof of identity. Finally, providing the synthetic peptide has an unprotected α-amino function, amino acid sequencing will provide absolute proof of its identity.

6. Synthesis of phosphorylated peptides

The protein kinases and phosphatases are known to play a pivotal role in the regulation of cellular proliferation, modulation of the cell cycle, and elaboration of the transformed phenotype in a variety of cell lines (72–75). In the light of this, much effort has been directed towards the synthesis of peptides containing phosphorylated serine, threonine, and tyrosine residues, for use in kinetic and immunological studies on these important classes of enzyme.

The most successful approaches to the synthesis of such peptides adhere to the following general scheme:

(a) Peptide chain elaboration is achieved using standard Fmoc SPPS protocols.

(b) Phosphitylation of the appropriate hydroxy-amino acid is then carried out whilst the peptide is still attached to the solid support.

(c) Oxidation to the phosphonate derivative is carried on the resin-bound peptide.

(d) Cleavage of the peptide from the support and removal of side-chain protecting groups, including those masking the phosphonate grouping.

The syntheses of the phosphonopeptides reported so far have been achieved using almost the whole array of standard coupling methodologies including Opfp active esters, HBTU, and BOP chemistries. Indeed, the only deviation from the standard Fmoc SPPS protocols that are normally employed is that the serine, threonine, or tyrosine residues to be phosphorylated are in-corporated with their side-chain hydroxyl functions unprotected (76–79). The activation of the α-carboxyl groups of Fmoc.Ser.OH, Fmoc.Thr.OH and Fmoc.Tyr.OH, using HBTU or the closely related derivative 2-(1H-benzotriazole-1-yl)-1,1,3,3-tetramethyluronium tetrafluoroborate, is reported to proceed very smoothly, essentially using the method outlined in *Protocol 13* (77, 78).

The preferred methods for the phosphitylation of the hydroxyl-containing amino acids on the resin-bound peptide, involve the use of either dimethyl- or di-*t*-butyl-*N*,*N*,-diisopropylphosphoramidite (76, 78–81). This will result in the generation of phosphitylated peptides protected at phosphorus with either methyl or *t*-butyl esters respectively. The details for this procedure, the subsequent oxidation to the phosphonate derivative, and the cleavage of the peptide from the resin and side-chain protecting group removal are given in *Protocol 22*.

Protocol 22. Synthesis of phosphonopeptides using 'solid-phase phosphorylation'

Reagents

- Dimethyl or di-*t*-butyl-*N*,*N*-diisopropyl-phosphoramidite
- Dimethylacetamide (DMA)
- *t*-Butylhydroperoxide
- Tetrazole
- DMF
- DCM

Method

1. Synthesize the peptide using the standard deprotection and coupling methods outlined in *Protocols 11–13*. Any of the functionalized resins listed in *Figures 8* and *9* can be utilized for this purpose, the exact choice obviously being dependent on whether a peptide amide or acid is required. The serine, threonine or tyrosine residues to be phos-phorylated are incorporated with their OH functions unprotected. Use *Protocol 13* for the activation of these amino acids.

2. Phosphitylate the resin-bound peptide by reaction with a solution of either dimethyl- or di-*t*-butyl-*N*,*N*-diisopropylphosphoramidite (10–20 equiv based on resin loading) in DMA (10–20 ml/g of resin). This is carried out in the presence of DNA-synthesis grade tetrazole (35 equiv, this reagent is required to activate the phosphoramidite). The reaction is allowed to proceed for 20–60 min, at room temperature.

3. Oxidize the phosphitylated peptide with *t*-butylhydroperoxide (82) (20 equiv, 10–30 min) dissolved in DMA.

4. Wash the resin with DMF (3 × 20 ml, 3 min per wash) followed by DCM (3 × 20 ml, 3 min per wash).

Protocol 22. *Continued*

5. If di-*t*-butyl-*N*,*N*-diisopropylphosphoramidite has been used to incorporate the phosphorus moiety, cleave the peptide from the solid support and remove all of the side-chain protecting groups (including the phosphonate *t*-butyl esters), concomitantly, by following one of *Protocols 17–20*. However, if the dimethyl analogue has been used, the TFA-resistant methyl esters from the resultant phosphonate moiety need to be cleaved by TFMSA. This can be achieved by carrying out an additional cleavage step, essentially by following *Protocol 10*.

References

1. du Vigneaud, V., Ressler, C., and Trippett, S. (1953). *J. Biol. Chem.*, **205**, 949.
2. du Vigneaud, V., Ressler, C., Swann, J. M., Roberts, C. W., and Katsoyannis, P. G. (1954). *J. Am. Chem. Soc.*, **76**, 3115.
3. Merrifield, R. B. (1963). *J. Am. Chem. Soc.*, **85**, 2149.
4. Carpino, L. A. and Han, G. Y. (1972). *J. Org. Chem.*, **37**, 3404.
5. Chang, C. D. and Meienhofer, J. (1978). *Int. J. Pept. Prot. Res.*, **11**, 246.
6. Meienhofer, J., Waki, M., Heimer, E. P., Lambros, T. J., Makofske, R. C., and Chang, C. D. (1979). *Int. J. Pept. Prot. Res.*, **13**, 35.
7. Barany, G., Kneib-Cordonier, N., and Mullen, D. G. (1987). *Int. J. Pept. Prot. Res.*, **30**, 705.
8. Fields, G. B. and Noble, R. L. (1990). *Int. J. Pept. Prot. Res.*, **35**, 161.
9. Merrifield, R. B. (1988). *Makromol. Chem. Makromol. Symp.*, **19**, 31.
10. Mitchell, A. R., Erickson, B. W., Ryabstev, M. N., Hodges, R. S., and Merrifield, R. B. (1976). *J. Am. Chem. Soc.*, **98**, 7357
11. Stewart, J. M. and Young, J. D. (1984). In *Solid phase peptide synthesis*, p. 101. Pierce Chemical Co., Rockford, Illinois.
12. DeGrado, W. F. and Kaiser, E. T. (1980). *J. Org. Chem.*, **45**, 1295.
13. Gesin, B. F. (1973). *Helv. Chim. Acta*, **56**, 1476.
14. Castro, B., Dormoy, J. R., Evin, G., and Selve, C. (1975). *Tetrahedron Lett.*, **16**, 1219.
15. Knorr, R., Trzeciak, A., Bannwarth, W., and Gillessen, D. (1989). *Tetrahedron Lett.*, **30**, 1927.
16. Stewart, J. M. and Young, J. D. (1984). In *Solid phase peptide synthesis*, p. 98. Pierce Chemical Co., Rockford, Illinois.
17. Tam, J. P., Heath, W. F., and Merrifield, R. B. (1983). *J. Am. Chem. Soc.*, **105**, 6442.
18. Tam, J. P., Heath, W. F., and Merrifield, R. B. (1986). *J. Am. Chem. Soc.*, **108**, 5242.
19. Bergot, B. J., Noble, R. L., and Geiser, T. (1986). In *User Bulletin No. 16, Peptide synthesizer*. Applied Biosystems Inc.
20. Atherton, E., Fox, H., Harkiss, D., Logan, C. J., Sheppard, R. C., and Williams, B. J. (1978). *J. Chem. Soc. Chem. Commun.*, 537.
21. Sheppard, R. C. (1973). In *Peptides 1971* (ed. H. Nesvadba), p. 111. North Holland, Amsterdam.

22. Kemp, D. S. (1973). In *Peptides 1971* (ed. H. Nesvadba), p. 1. North Holland, Amsterdam.
23. Atherton, E., Brown, E., Sheppard, R. C., and Rosevear, A. (1981). *J. Chem. Soc. Chem., Commun.*, 336
24. Atherton, E., Logan, C. J., and Sheppard, R. C. (1981). *J. Chem. Soc. Perkin Trans. 1*, 538.
25. Sheppard, R. C. and Williams, B. J. (1982). *Int. J. Pept. Prot. Res.*, **20**, 451.
26. Atherton, E. and Sheppard, R. C. (1989). In *Solid phase peptide synthesis, a practical approach* (ed. E. Atherton and R. C. Sheppard), p. 63. IRL Press, Oxford.
27. Dryland, A. and Sheppard, R. C. (1986). *J. Chem. Soc. Perkin Trans. 1*, 125.
28. Wang, S-S. (1973). *J. Am. Chem. Soc.*, **95**, 1328.
29. Atherton, E., Brown, E., Priestly, G., Sheppard, R. C., and Williams, B. J. (1981). In *Proceedings of the 7th American Peptide Symposium* (ed. D. H. Rich and E. Gross), p. 163. Pierce Chemical Co., Rockford, Illinois.
30. Rink, H. (1987). *Tetrahedron Lett.*, **28**, 3787.
31. Mergler, M., Tanner, R., Gosteli, J., and Grogg, P. (1988). *Tetrahedron Lett.*, **29**, 4005.
32. Albericio, F., Kneib-Cordonier, N., Biancalana, S., Gera, L., Masada, R. I., Hudson, D., and Barany, G. (1990). *J. Org. Chem.*, **55**, 3730.
33. Sheppard, R. C. (1983). *Chem. Br.*, **19**, 402.
34. Atherton, E., Cameron, L., Meldal, M., and Sheppard, R. C. (1986). *J. Chem. Soc. Chem. Commun.*, 1763.
35. Dryland, A. and Sheppard, R. C. (1988). *Tetrahedron*, **44**, 859.
36. Small, P. W. and Sherrington, D. C. (1989). *J. Chem. Soc. Chem. Commun.*, 270.
37. Bayer, E. and Rapp, W. (1986). *Chem. Pept. Prot.*, **3**, 3.
38. Bayer, E. (1991). *Angew. Chemie*, **103**, 117.
39. Kirstgen, R., Sheppard, R. C., and Steglich, W. (1987). *J. Chem. Soc. Chem. Commun.*, 1870.
40. Kirstgen, R. and Steglich, W. (1989). In *Peptides 1988* (ed. G. Jung and E. Bayer), p. 148. Walter de Gruyter, Berlin.
41. Eberle, A. N., Atherton, E., Dryland, A., and Sheppard, R. C. (1986). *J. Chem. Soc. Perkin Trans. 1*, 361.
42. Atherton, E., Dryland, A., Goddard, P., Cameron, L., Richards, J. D., and Sheppard, R. C. (1988). In *Peptides: structure and function* (ed. C. M. Deber, V. J. Hruby, and K. D. Kopple), p. 249. Pierce Chemical Co., Rockford, Illinois.
43. Fujino, M., Wakimasu, M., and Kitada, C. (1981). *Chem. Pharm. Bull. Japan*, **29**, 2825.
44. Atherton, E., Cammish, L. E., Goddard, P., Richards, J. D., and Sheppard, R. C. (1985). In *Peptides 1984* (ed. U. Ragnarsson), p. 153. Almqvist and Wiksell International, Stockholm.
45. Atherton, E., Sheppard, R. C., and Wade, J. D. (1983). *J. Chem. Soc. Chem. Commun.*, 1060.
46. Green, J., Ogunjobi, O. M., Ramage, R., Stewart, A. S. J., McCurdy, S., and Noble, R. (1988). *Tetrahedron Lett.*, **29**, 4341.
47. Atherton, E., Woolley, V., and Sheppard, R. C. (1980). *J. Chem. Soc. Chem. Commun.*, 970.
48. Tam, J. P. (1988). *Proc. Natl. Acad. Sci. USA*, **85**, 5409.

49. Drijfhout, J. W., Freijlbrief, M., Welling, G. W., Welling-Wester, S., and Bloemhoff, W. (1989). In *Abstracts of the 11th American Peptide Symposium*, p. 254.

50. Gish, D. T., Katsoyannis, P. G., Hess, G. P., and Stedman, R. J. (1956). *J. Am. Chem. Soc.*, **78**, 5954.

51. Wegand, F., Steglich, W., Bjarnason, J., Ahktar, R., and Chytill, N. (1968). *Chem. Ber.*, **101**, 3623.

52. Konig, W. and Geiger, R. (1970). *Chem. Ber.*, **103**, 2041.

53. Sieber, P. and Riniker, B. (1990). In *Innovations and perspectives in solid phase synthesis 1990, 1st international symposium proceedings* (ed. R. Epton), p. 577. SPPC, Birmingham.

54. Brown, T., Jones, J. H., and Richards, J. D. (1982). *J. Chem. Soc. Perkin Trans. 1*, 1553.

55. Colombo, R., Colombo, F., and Jones, J. H. (1984). *J. Chem. Soc. Perkin Trans. 1*, 292.

56. Seyer, R., Aumelas, A., Caraty, A., Rivaille, P., and Castro, B. (1990). *Int. J. Pept. Prot. Res.*, **35**, 465.

57. Gairi, M., Lloyd-Williams, P., Albericio, F., and Giralt, E. (1990). *Tetrahedron Lett.*, **31**, 7363.

58. Fields, C. G., Lloyd, D. H., Macdonald, R. L., Ottesen, K. M., and Noble, R. L. (1991). *Pept. Res.*, **4**, 95.

59. Martinez, J., Bali, J-P., Rodriquez, M., Casro, B., Laur, J., and Lignon, M-F. (1988). *J. Med. Chem.*, **28**, 1874.

60. Kaiser, E., Colescott, R. L., Bossinger, C. D., and Cook, P. I. (1970). *Anal. Biochem.*, **71**, 261.

61. Hancock, W. S. and Battersby, J. E. (1976). *Anal Biochem.*, **84**, 595.

62. Kaiser, E., Bossinger, C. D., Colescott, R. L., and Olser, D. D. (1980). *Anal. Chem. Acta*, **118**, 149.

63. McFerran, N. V. and Walker, B. (1990). In *Innovations and perspectives in solid phase synthesis 1990, 1st international symposium proceedings* (ed. R. Epton), p. 261. SPPC, Birmingham.

64. McFerran, N. V., Walker, B., McGurk, C. D., and Scott, F. (1991). *Int. J. Pept. Prot. Res.*, **37**, 382.

65. Nielsen, C-S., Hansen, P. H., Lihme, A., and Heegard, P. M. H. (1990). In *Innovations and perspectives in solid phase synthesis 1990, 1st international symposium proceedings* (ed. R. Epton), p. 549. SPPC, Birmingham.

66. King, D. S., Fields, C. G., and Fields, G. B. (1990). *Int. J. Pept. Prot. Res.*, **36**, 255.

67. Sieber, P., Kamber, B., Riniker, B., and Rittel, W. (1980). *Helv. Chim. Acta*, **63**, 2358.

68. Otaka, A., Koide, T., Shide, A., and Fujii, N. (1991). *Tetrahedron Lett.*, **32**, 1223.

69. Tam, J. P., Wu, C-R., Liu, W., and Zhang, J-W. (1991). *J. Am. Chem. Soc.*, **113**, 6657.

70. Rivier, J., McClintock, R., Galyean, R., and Anderson, H. (1984). *J. Chromatogr.*, **288**, 303.

71. Lloyd-Williams, P., Albericio, F., and Giralt, E. (1991). *Int. J. Pept. Prot. Res.*, **37**, 58.

72. Yarden, Y. and Ulcer, A. A. (1988). *Ann. Rev. Biochem.*, **57**, 443.
73. Gould, K. and Nurse, P. (1989). *Nature*, **342**, 39.
74. Rosen, O. M. (1987). *Science*, **237**, 1452.
75. Heldin, C-H., Betsholtz, C., Claesson-Welch, L., and Westmark, B. (1987). *Biochim. Biophys. Acta*, **907**, 219.
76. Andrews, D. M., Kitchin, J., and Seale, P. W. (1991). *Int. J. Pept. Prot. Res.*, **38**, 469.
77. Trzeciak, A., Vorherr, T., and Bannwarth, W. (1992). *22nd European Peptide Symposium (Interlaken, Switzerland)*, abstract No. **P89**.
78. Kitas, E. A., Knorr, Trzeciak, A., and Bannwarth, W. (1991). *Helv. Chim. Acta*, **74**, 1314.
79. Bannwarth, W. and Trzeciak, A. (1987). *Helv. Chim. Acta*, **70**, 175.
80. Perich, J. W., Le Nguyen, D., and Reynolds, E. C. (1991). *Tetrahedron Lett.*, **32**, 4033.
81. Perich, J. W. and Johns, R. B. (1990). *Australian J. Chem.*, **43**, 1623.
82. Ottinger, E. A., Sole, N. A., Tian, Z., Bernlohr, D. A., and Barany, G. (1993). In *Peptides 1992* (ed. C. H. Schneider and A. N. Eberle), p. 334. ESCOM Science Publishers, Leiden.

4

Immunization with peptide–carrier complexes: traditional and multiple-antigen peptide systems

JAMES P. TAM

1. Introduction

Among various strategies for immunization, the strategy of using synthetic peptides is perhaps the most appealing, both in conceptual simplicity and in practical convenience (1–10). It is a minimalist approach that uses a peptide segment with one or a few of the desired epitopes of a protein or whole organism, both of which are many times larger than the peptide. By keeping only the essential epitope(s) and eliminating all other, unwanted, portions of the protein or organism, a synthetic peptide immunogen provides, in theory, the desired selectivity. Advances in solid-phase synthesis have made the preparation of peptides a routine task in most laboratories. In addition, access to synthetic peptides has been made easy by the proliferation of core facilities in universities and of commercial laboratories specialized for the preparation of peptides to order. Thus, the use of synthetic peptides has become a well accepted method for the preparation of site-specific antibodies for various biochemical and functional studies.

Peptides are generally poor immunogens and are cleared from the body rapidly. For immunization purposes, peptides have to be modified and administered with a formulation containing an adjuvant and a vehicle such as paraffin oil. In this chapter, I will describe the different approaches for peptide modification and presentation. These will include the traditional approaches of conjugation to a protein carrier and polymerization, as well as the new approach known as the multiple-antigen peptide (MAP) which contains well defined structures with or without a built-in adjuvant. The chapter is organized into four sections. First, the criteria for peptide immunogenicity are described. Second, the traditional approaches of peptide protein–carrier conjugates and polymerization are discussed. In the next section a thorough overview of the new MAP approach is given. The final section provides practical information concerning immunization and the preparation, storage, and purification of antibodies from sera and ascitic fluids.

2. Traditional and MAP approaches

Protein carriers prevent the rapid degradation or clearance of the attached peptide and enhance its immunogenicity. In addition, protein carriers may have the ability to induce the desired conformation in the peptide epitope and to provide a source of T-helper cell epitopes. For these reasons, the traditional methods using peptide–protein conjugates have a proven record of success. They also have limitations. A major limitation is that the large size of the protein carrier, as compared with the peptide, means that the peptide epitope represents only a minor fraction of the total peptide–carrier protein conjugate. Thus, the desired antibodies may represent only a minor fraction of the total number of antibodies produced. In this regard, conjugation to a protein carrier appears to contradict the aim of simplicity in a peptide-based immunogen. In terms of vaccines for humans, the undesirable features attendant on the use of a protein carrier include the creation of many irrelevant epitopes that cause carrier toxicity (9) and epitopic suppression (10). In large-scale production, peptide conjugation to a protein carrier may have the problem of batch-to-batch inconsistency due to the chemical ambiguity of the peptide–protein carrier structure. Also, the complexity of attaching peptide antigens to large proteins limits the incorporation of more than one epitope to the protein carrier.

A new technique, recently developed in the author's laboratory (11–24), is referred to as the MAP approach. The MAP approach replaces the protein carrier with a small core matrix comprising oligomeric lysine; it is a unique presentation system that provides peptide epitopes unambiguously in multiple copies. As a result, a MAP provides a very high density of the desired peptide epitopes at the surface of the construct with a small non-protein core matrix (<800 Da) as a scaffold. This eliminates many of the disadvantages associated with the use of a protein carrier.

For the preparation of peptide immunogens, the steps are slightly different in the traditional and MAP approaches. The following steps are necessary in the traditional approach to prepare site-specific antibody:

(a) Select a peptide sequence as an immunogen.

(b) Select a carrier.

(c) Select a conjugation method to prepare the peptide–carrier complex.

(d) Deliver the peptide–carrier in a formulation for immunization.

In the MAP approach steps (b) and (c) are not required. Both approaches will be discussed separately below.

2.1 Selection of peptides

Common to all of the peptide-based approaches is the selection of a peptide sequence as an immunogen. Two criteria are considered in peptide selection: immunogenicity and antigenicity. Immunogenicity is the ability of the peptide to elicit high-titred antibody. Antigenicity is the ability of the antibody to recognize the peptide and the protein from which it is derived; this is governed by the conformation of the peptide. In simple terms, one must choose a peptide sequence for the immune system to respond to and provide the correct conformation so that the ensuing antibody will recognize the parent protein. In selecting a peptide, the correct conformation is clearly important because there are methods by which to increase the immunogenicity of the peptide, as shown in the following sections.

2.2 B- and T-cell epitopes

The ability of mammalian hosts to produce a vigorous immune response to foreign immunogens requires the cooperation of B- and T-cells. B-cells function through the production of antibodies and are the major architect of the humoral response. T-cells function through the production of various cytokines and immunoregulators and are the vehicles of the cell-mediated response. A given peptide or protein is processed by immune cells and is used to activate the B- and T-cells. The minimal length of a continuous polypeptide sequence that can activate these cells is referred to as either a B- or T-cell epitope. B-cell epitopes can be either continuous or discontinuous while only continuous T-cell epitopes are known to date.

Structurally, B-cell epitopes are primarily defined on the basis of accessibility of regions of the native protein to react with antibodies. Based on the results obtained from their three-dimensional structures, all antigen–antibody complexes show shape and chemical complementarity of their interacting surfaces. Almost any peptide can be a B-cell epitope. However, most peptide epitopes are selected because they represent a structural characteristic of the parent protein. The best choices are offered by sequences that are located on the surface, near the termini, or at reverse turns (see Chapter 2).

Experience has shown that T-cell epitopes are difficult to define and contain several categories. For the production of antibodies a T-helper cell epitope is necessary. Identification of these epitopes can be accomplished by a systematic mapping using a panel of sequences covering a stretch of a protein, followed by sequential testing of each peptide. An alternative approach to identifying T-cell epitopes is based on predictive computer modelling programs (see Chapter 2).

The minimal length of a continuous B-cell epitope is controversial. However, from the results of epitope scanning, a length of six amino acids is usually agreed to be functionally minimal. T-helper cell epitopes have been shown to be of 9–11 amino acids, but longer lengths have also been found.

B- and T-helper cell epitopes are not necessarily exclusive of each other. Thus, a longer peptide (about 15 amino acids) may contain overlapping B- and T-helper cell epitopes. In using the MAP approach, it is best to use a longer peptide and place the B-cell epitope near the N-terminus or a reverse turn region to optimize the production of high-affinity antibodies.

Factors influencing the kinetics and quantity of an antibody response by the host animal are largely controlled by the antigen itself and by its ability to activate cell-to-cell communication between B-cells and T-helper cells. The B- and T-cell contact is mediated by the antigen by its binding to both the major histocompatibility complex (MHC) class II protein and the T-cell receptor. Thus, a good immunogen should generally be able to activate both B- and T-cells. The requirements for activating both B- and T-cells can be met in the following ways:

(a) coupling the peptide to a protein carrier which provides T-helper cell epitopes

(b) coupling the peptide to a peptide which is a known T-cell epitope

(c) coupling the peptide to a lipid or a macroparticular polymer such as agarose which is taken up by antigen-presenting cells

(d) co-administering the peptide with a polyclonal immune cell activator such as Freund's complete adjuvant or lipid A

3. Traditional approaches

3.1 Types of protein carrier

To prepare site-specific antibody by the traditional approach, two decisions have to be made: the choice of a carrier and the method of peptide–carrier conjugation. Because most protein carriers are readily available and have a proven record of success, proteins are widely used. Polymeric carriers such as polyethylene glycol and polyamino acids are less immunogenic and are sometimes used to protect a peptide drug against degradation and to minimize rather than to enhance antibody responses.

Nearly any protein can serve as a carrier and the list includes: invertebrate proteins such as keyhole limpet haemocyanin (KLH), albumins such as bovine serum albumin (BSA) and ovalbumin, deactivated bacterial toxins such as tetanus toxoid and diphtheria toxoid, and other large and small proteins such as fowl immunoglobulin, ribonuclease, myoglobin, and thyroglobulin. The criteria for choosing a protein carrier include: solubility, sufficient reactive side-chains for coupling to the synthetic peptide, and minimal cross-reactivity with other proteins. Smaller proteins such as ribonuclease and myoglobin have the disadvantage of limited numbers of conjugation sites. Antisera derived from albumins have the obvious limitation of cross-reactivity with tissue culture media and they give high backgrounds when albumin is used to block plastic surfaces in enzyme-linked immunosorbent assay (ELISA). When thyroglobulin is used as carrier the anti-thyroglobulin produced may react non-specifically with iodinated proteins and KLH becomes insoluble when excessive cross-linking occurs with glutaraldehyde. Nevertheless, in surveying the literature, KLH and BSA appear to be the most commonly used carriers in the laboratory. Tetanus and diphtheria toxoid which are potent immunogens are the preferred choices as carriers used for vaccines. Whatever the choice, the conjugation of the peptide to the carrier is probably the most often overlooked aspect and it causes many failures in the production of antibodies.

3.2 Conjugation and polymerization methods

The conjugation of a peptide to a protein carrier can be accomplished by one of the many commonly used reagents listed in *Table 1*. In general, there are two broad categories: non-specific, random conjugation and site-specific, random conjugation

The non-specific, random methods rely on cross-linking in a single step to join carboxylic (aspartate or glutamate), sulphydryl (cysteine), amino (lysine), or phenolic (tyrosine) groups. Cross-linking reagents such as 1-ethyl-3-(3-dimethylaminopropyl)-carbodiimide (EDC), glutaraldehyde and bis-diazotized benzidine (BDB) are used in one-stage conjugation methods in which the reagent, peptide, and protein are mixed to effect the reaction. Obviously, the disadvantage of this approach is that the epitope on the peptide may be modified, and self-linking of the peptide or protein to itself can and will occur.

These reagents are also used for the polymerization of peptide epitopes. To avoid the modification of the epitopic sites, a general approach is to place two cysteines, commonly one at each terminal, to effect polymerization via disulphide bond formation. A limitation of all the polymerization approaches is that there is little control on the size of the polymers obtained. Recently, two or three peptide antigens have been used in polymerization to allow multiple incorporation of two or more epitopes (multiple T- and B-cell epitopes) to effect a broad range of immune responses (21). A case in point is the human malaria vaccine currently under clinical trials which is produced by polymerization of three peptide antigens containing cysteines in their termini.

To avoid self-linking, a series of heterobifunctional cross-linking reagents has been developed (25–31). These site-specific, random reagents (*Table 1*) require a two-stage reaction. A popular and effective reagent is 3-maleimidobenzoic acid N-hydroxysuccinimide ester (MBS). Its maleimido moiety reacts with a sulphydryl on the peptide first and then the derivatized peptide reacts randomly with amino groups in the protein carrier via the active ester of hydroxysuccimide. The first reaction of the peptide with the cross-linking reagent usually determines the orientation of the peptide

Table 1. Cross-linking reagents for preparing peptide–protein conjugates. The reagents are defined as follows: EDC, 1-ethyl-3-(3-dimethylpropyl)-carbodiimide; Glut, glutaraldehyde; BDB, bis-diazotized benzidine; MBS, 3-maleimidobenzoic acid N-hydroxysuccinimide ester; SPDP, N-succinimidyl-3-(2-pyridylthio)propionate; SATA, N-succinimidyl-S-acetylthioacetate; SMPT,4-succinimidyloxycarbonyl-α-methyl-α-(2-pyridyldithio)-toluene; SMCC, N-succinimidyl 4-(N-maleimidomethyl)cyclohexane-1-carboxylate; SIAC, N-succinimidiyl (4-idoacetyl)cyclohexane-1-carboxylate; SAB, N-succinimidyl-4-azidobenzoate

Name	Structure	Reaction	Product
Non-specific, random			
EDC	$R-C=N=C-R'$	$NH_2/COOH$	Amide
Glut	$OHC-CH_2-CH_2-CHO$	NH_2/NH_2	Schiff base
BDB		Tyr/Tyr	Azo-compound
Site-specific, random			
MBS		NH_2/SH	S-R/amide
SPDP		NH_2/SH	S-S/amide
SATA		NH_2/SH	S-S/amide
SMPT		NH_2/SH	S-S/amide
SMCC		NH_2/SH	S-R/amide
SIAC		NH_2/SH	S-R/amide
SAB		NH_2/XH	X-R/amide

linkage to the the carrier because, for example, a cysteine can be placed at either the C- or N-terminal of the peptide. The choice of orientation and cross-linking reagent is guided by the sequence and the position of the peptide in the native protein.

3.3 Do's and don'ts

A criterion for conjugation is to preserve the integrity of the peptide epitope. The following simple rules apply for the non-specific, random reagents in these procedures. *Do not* use the coupling reagent if the following internal residues are found in the peptide: BDB: tyrosine, EDC: glutamate, aspartate or lysine, and glutaraldehyde: lysine or arginine. *Do not* use amine buffers such as Tris–HCl for most conjugation reactions; use phosphate buffer instead. Similarly, *do not* use acetate buffer for EDC conjugation.

Protocol 1 provides a method of coupling a peptide to BSA using glutaraldehyde. Typically the degree of incorporation of peptide ranges from 40 to 80%. This may be determined by adding a small amount of labelled peptide (^{125}I is usually used) with the cold peptide during the conjugation. The incorporation is measured by counting the trichloroacetic acid-precipitated conjugate (32).

The coupling of a cysteine-containing peptide to KLH using the heterobifunctional reagent MBS is described in *Protocol 2*. The incorporation of peptide is usually in the range 40 to 60%. Another procedure for coupling a thiol-containing peptide to a protein is given in Chapter 6, *Protocol 7*.

Protocol 1. Conjugation of peptide to BSA using glutaraldehyde[a]

Reagents

- BSA
- 0.2 M sodium borate buffer, pH 8.5
- Glutaraldehyde (22 mM, freshly prepared)
- Ethanolamine
- Phosphate-buffered saline (PBS), 20 mM sodium phosphate buffer, pH 7.2, 0.15 M NaCl

Method

1. Mix the peptide[b] with BSA in borate buffer to final concentrations of 0.1 mg/ml and 6 mg/ml respectively; for a 10-mer this gives a molar ratio of 10:1.

2. Add glutaraldehyde dropwise to the solution, at room temperature, to a final concentration of 5 mM.

3. Agitate the mixture overnight at room temperature using a mixing rocker or a stirring bar.

4. Add ethanolamine to a final concentration of 0.2 M and incubate for 30 min to quench unreacted aldehyde.

Protocol 1. *Continued*

5. Dialyse the mixture, using a membrane with an exclusion limit of 5000, overnight at 4 °C, against three changes of PBS.

---

[a] Modified from Beinfield and O'Donohue (32).

[b] Other amino compounds, such as Tris or glycine, must be removed as they will interfere with the reaction.

Protocol 2. Conjugation of a thiol-containing peptide to KLH using MBS[a,b]

Reagents

- 0.1 M sodium phosphate buffer, pH 7.0
- MBS (25 mg/1 ml of dimethyl formamide) (Pierce Chemical Co.)
- 2-Mercaptoethanol
- KLH (Pierce Chemical Co.)
- PBS (see *Protocol 1*)

Method

1. Dissolve the peptide (5 mg/ml) in the phosphate buffer.[c]

2. Add MBS dropwise with stirring to give a final concentration of 1 mg/ml and continue stirring for 30 min at room temperature.

3. Add 2-mercaptoethanol to a final concentration of 35 mM and incubate for 30 min at room temperature to quench unreacted maleimide groups.

4. Dissolve sufficient KLH in PBS in a small reaction vessel to give a KLH:peptide molar ratio of 1:40.[d]

5. Add the activated peptide dropwise with stirring to the KLH solution and continue stirring for 3 h.

6. Dialyse the mixture overnight at 4 °C against three changes of PBS.

---

[a] Modified from Kitagawa and Aikawa (28).

[b] In an alternative method KLH is first activated with MBS and the activated KLH is used to couple the thiol-containing peptide. This approach may lead to a small amount of disulphide exchange but has been found to be as effective as the method described. It has the advantage that excess MBS can be removed by gel filtration or dialysis prior to the coupling to the peptide.

[c] 2-mercaptoethanol and other reducing agents (e.g. dithiothreitol) must be removed prior to coupling because they interfere with the reaction.

[d] KLH and its conjugates are bluish and opalescent. They are known to be poorly soluble in water which makes purification by gel permeation difficult.

4. The MAP approach

4.1 The MAP concept

The inherent disadvantages of using a protein carrier or peptide polymer-ization can be resolved by the MAP approach; this replaces large protein carriers with a small structural unit that carries multiple copies of the peptide antigen in a controlled and unambiguous way. The conceptual framework of the MAP approach is based on the design of a small core matrix with the following properties:

- lack of immunogenicity
- presentation of multiple copies of the epitope
- ability to incorporate different epitopes
- ease of chemical synthesis

A core matrix consisting of branched trifunctional amino acids such as lysine has satisfied these requirements (11).

Since lysine contains two amino groups, each level (n) of propagation will contain $2n$-1 molecules of lysine and $2n$ of reactive amino ends. Thus, the first level of the core matrix, consisting of one lysine, contains two amino groups; the second level, consisting of three lysines, contains four amino groups; the third level, consisting of seven lysines, contains eight amino groups, and so on (*Figure 1*). Usually, the second or third level of propagation is sufficient to give a low molecular weight core matrix. Higher levels of propagation have not shown any advantages over those that contain two or three levels of branching. Peptides are attached to the amino groups of the lysine scaffold to give a macromolecule that has a high density of uniformly distributed peptide antigens on the surface. To illustrate this point, a 20-mer derived from the sequence of the foot-and-mouth disease virus (FMDV) protein VP1 residues 141–160 attached to an octameric (three levels) branched core matrix consist-ing of seven lysines will give a MAP of about 23 000 Da. The weight due to eight copies of the peptide accounts for 96% of the total, and the core represented by seven lysines accounts for only 4%. In contrast, the same peptide when conjugated to a protein carrier such as KLH (molecular weight greater than 10^6) will give a low density of peptide epitope randomly dis-tributed on the protein carrier surface.

It is important to emphasize that the key component, the core matrix, is composed of an oligomeric branched lysine rather than polymeric lysine, and it has a molecular weight of less than 900 Da. Polylysine is usually a large protein with a molecular weight greater than 10^5 and it contains many side-chain cationic groups. In contrast, the core matrix does not carry any cationic charges since all the side-chain amino groups are coupled as amide bonds either to another lysine or to peptide.

From the practical point of view, the MAP approach for the preparation of

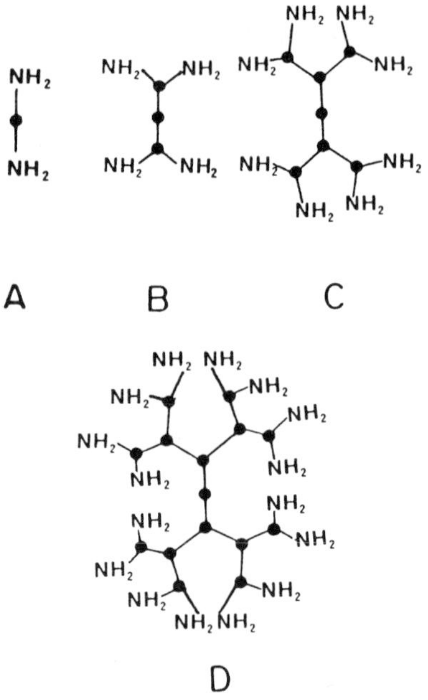

Figure 1. Schematic representation of the core matrix of the MAP. A: first level, divalent; B: second level, tetravalent; C: third level, octavalent; D: fourth level, decavalent.

immunogens has several advantages. First, it can be synthesized *de novo* in an unambiguous manner. Thus, both the structure and stoichiometric ratio of peptide to the core matrix is unambiguous. Unlike the conventional peptide–carrier conjugates, the MAP structure can be unequivocally represented by a chemical formula and can be verified with great precision by analytical methods such as amino acid analysis, sequencing, and mass spectrometry; this property is essential for quality control and consistent batch-to-batch production. Second, there is no need for conjugation to a protein carrier which may cause such undesirable immunological reactions as epitopic suppression. Finally, there is the advantage of flexibility that allows the design and engineering of both B- and T-cell epitopes as well as a built-in adjuvant. This flexibility is due to the sophistication of peptide synthetic methodology, which allows the α- and the side-chain amino groups of lysine to be distinguished chemically and different epitopes to be attached selectively. The inclusion of T-cell epitopes that could enhance the immunogenicity or elicit cell-mediated immune responses is an important consideration in the design of vaccines. In short, the MAP concept provides many of the desirable features for the preparation of site-specific antibodies for laboratory uses as well as for the design of the peptide-based vaccines.

MAPs also have advantages in solid-phase immunoassay of anti-peptide antibodies. When single-chain peptides are placed on solid phases they may not bind or, if they do, part of the epitope may not be available for binding to the antibody or the peptide may be denatured. A MAP has multiple copies of the peptide available for both binding to surfaces and to antibodies. This leads to increased and reproducible immobilization of the antigen.

4.2 The MAP strategy

4.2.1 Preparation of MAPs

For the preparation of MAPs, two general strategies are available: the direct and the indirect or modular approaches. The direct approach (11, 12) is simpler in execution than the indirect approach (24) since it is a stepwise preparation of the core matrix and the peptide immunogen in a single, continuous operation using solid-phase peptide synthesis. The indirect approach has the advantage that purified synthetic peptides are used. The chemistry for both approaches is adopted from the conventional solid-phase peptide methodology (33–35). Both methods are no more complicated than the conventional method for the preparation of a monomeric peptide. Before venturing into the specific details of MAP synthesis, several comments on the preparation of peptide antigens for use in MAPs are warranted.

4.2.2 Background on solid-phase peptide synthesis

Most solid-phase peptide synthesis follows one of two procedures. Both procedures involve sequential peptide synthesis whereby a single amino acid is successively added at each step, starting from the C-terminus of the peptide. They differ in the protecting group strategy which governs the protection and cleavage of the α-amino and side-chain groups as well as the method used for deprotection and cleavage from the solid support.

The older and more commonly used procedure is *t*-butyloxycarbonyl (Boc) chemistry which is based on the protecting group strategy of differential acid lability. The α-amino group is protected by a Boc-protecting group which requires a mild acid, such as trifluoroacetic acid (TFA), for its removal. The side-chain is protected with a benzyl alcohol derivative and requires a strong acid, such as HF, to remove all the benzyl-type protecting groups and to cleave the peptide from the resin. Thus, the operating principle for Boc chemistry is the exploitation of differential acid strengths for the removal of protecting groups.

The second procedure is 9-fluorenylmethyloxycarbonyl (Fmoc) chemistry. The Fmoc group is used for the protection of the α-amino group and requires an organic base, such as piperidine, for its removal. The side-chain protecting groups use the *t*-butyl alcohol derivatives and thus can be removed by TFA. Because base and acid are used on a different set of protecting groups, the Fmoc chemistry relies on the principle of orthogonality for the removal of protecting groups. The Boc method usually uses HF which requires special apparatus and a designated hood for its handling, thus Fmoc chemistry is appealing to many new users of peptide chemistry for its convenience. However, alternative deprotecting agents such as trifluoromethane sulphonic acid (TFMSA), HBr/TFA/pentamethylbenzene, have been developed and have started to gain support in several laboratories. These reagents do not require special apparatus and are relatively simple to handle. Nevertheless, several comparative studies, including those performed in this author's laboratory, have shown both Fmoc and Boc chemistries to be effective. In general, many side-reactions in the Boc chemistry are better understood and the reagents for Boc chemistry are less expensive than those for Fmoc chemistry.

Amino acids can be attached to the resin, or resins can be purchased with the desired amino acid already attached and synthesis can be performed either manually or by an automated peptide synthesizer.

4.2.3 Synthesis of MAPs by the direct approach

In the direct approach, a peptide antigen and the lysine core matrix are synthesized as a single unit on a resin support (*Figure 2*). The process begins with a resin support containing a simple amino acid such as β-alanine, which serves as an internal standard for calculating the molar ratio of other amino acids. The first level of the core matrix contains a diprotected lysine, usually Boc-Lys (Boc) for the Boc chemistry and Fmoc-Lys (Fmoc) for the Fmoc chemistry. Since the protection scheme of the α- and the ε-amino groups are similar, deblocking the Boc groups followed by coupling a new round of Boc-Lys (Boc) will furnish the next level of the core matrix containing two lysines, one at the α-amino position and the other at the ε-amino position. This

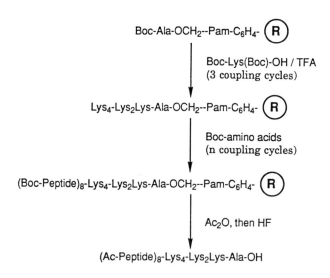

Figure 2. Synthesis of a MAP by the direct method.

second level of branching will give a tetrameric core containing four amino groups. Similarly, the third-level branching will add four lysines and eight amino groups, and give an octameric core. After the completion of the core matrix, the synthesis of the desired peptide epitope continues in the same reaction vessel; however, four or eight copies of the peptide are synthesized. At the completion of the synthesis, the peptide is usually capped with an acetyl group. The necessity for acetylation depends on the position of the peptide in the protein sequence. Capping is appropriate for a peptide derived from internal sequences since it would remove the charged amino group. The whole unit is then cleaved from the resin support by a strong acid. The product may be purified by dialysis or the usual chromatographic techniques (see Section 4.2.5), and can be used directly for immunization.

Protocol 3 describes the procedure for synthesizing MAPs by the direct approach using Boc chemistry. This procedure gives a 40–80% yield and 0.6 to 1 g of MAP containing an octomeric 10-residue peptide.

Protocol 3. Synthesis of MAPs by the direct method

Equipment and reagents

- Glass reaction vessel with coarse sintered glass filter at one end and a screw cap at the other (Rocky Mountain Scientific Glass-blowing)
- HF apparatus (Peptide International, Peninsula Laboratories)
- Boc-βAla-OCH$_2$-PAM resin (0.2 mmol/g) (Novobiochem)
- Boc-Lys(Boc) and other Boc-amino acids (Novabiochem)
- Dichloromethane (DCM)

- TFA (**Caution:** see note on p. 70)
- Diisopropylethylamine (DIPEA) (5% in TFA:DCM (1:1 v/v)) freshly prepared
- *N,N'*-Dicyclohexylcarbodiimide (DCC) (1 M in DCM)
- 1-Hydroxybenzotriazole (HOBt) (1 M in DMF)
- Dimethylformamide (DMF)
- 4-Dimethylaminopyridine (DMAP)
- Acetic anhydride (10% in DCM)

Method

Each stage involves a drain step and the addition of 15 ml of solvent unless specified.

Preliminary steps

1. Swell the Boc-βAla-OCH$_2$-PAM resin (0.2 mmol) in 15 ml of DCM for at least 2 h in a reaction vessel that holds 30 ml of solvent.

Cycle 1

2. Wash the resin for 1 min periods with DCM three times.
3. Remove the Boc group with TFA:DCM (1:1, v/v) first for 2 min and repeat with a 20 min agitation.
4. Wash the resin for 1 min periods with: DCM four times, 5% DIPEA two times, and DCM five times.

Protocol 3. *Continued*

5. Add Boc-Lys(Boc) (0.24 mmol, 4 equiv) in 5 ml of DCM and agitate the suspension for 1 min. **Do not drain** in this step or steps **6** and **7**.

6. Add DCC (0.24 ml, 0.24 mmol) followed by HOBt (0.24 ml, 0.24 mmol) and agitate the suspension for 15 min.

7. Add DMF (14 ml) and agitate for an additional 45 min.

8. Drain and wash the resin for 1 min periods with: DMF two times, and DCM two times.

9. Apply the ninhydrin test (Chapter 3, *Protocol 14*). If positive: repeat steps **5–8** with a different coupling reagent.[a] If negative: proceed to the next cycle.

Cycle 2

10. Repeat steps **2–4**.

11. Add Boc-Lys(Boc) (0.48 mmol) in 5 ml of DCM. **Do not drain** in this step or steps **12** and **13**.

12. Add DCC (0.48 ml, 0.48 mmol) followed by HOBt (0.48 ml, 0.48 mmol) and agitate the suspension for 15 min.

13. Add DMF (14 ml) and agitate for an additional 45 min.

14. Repeat steps **8** and **9**.[b]

Cycle 3

15. Repeat steps **2–4**.

16. Add Boc-Lys(Boc) (0.96 mmol) in 5 ml of DCM. **Do not drain** in this step or steps **17** and **18**.

17. Add DCC (0.96 ml, 0.96 mmol) followed by HOBt (0.96 ml, 0.96 mmol) and agitate the suspension for 15 min.

18. Add DMF (10 ml) and agitate for an additional 45 min.

19. Repeat steps **8** and **9**.[c]

Peptide extension, deprotection, and cleavage

20. Synthesize the required peptide on the core matrix by repeating steps **15–19** but use 1.92 mmol of each amino acid, DCC, and HOBt

21. Cap the N-terminal amino group with 10% acetic anhydride and 1% DMAP for 5 min.[d]

22. If necessary, remove the His(Dnp) and Boc groups (see Chapter 3, *Protocols 7* and *5* respectively).

23. Cleave the peptide from the resin using the 'Low–High HF' procedure (see Chapter 3, *Protocol 9*). (**Caution:** see pp. 41–2.)

24. Extract the peptide with 0.1 M Tris–glycine buffer, pH 8.2, 8 M urea, and then with water.

25. Dialyse the peptide against 0.1 M Tris–glycine buffer, pH 8.2, 8 M urea.

26. Purify the MAP by gel permeation with 2–5% acetic acid as solvent.

[a] Use either a phosphonium salt, such as benzotriazole-1-yl-oxy-tris-pyrrolidinophosphonium hexafluorophosphate, or a uronium salt, such as 2-(1-*H*-benzotriazole-1-yl-)-1,1,3,3-tetramethyluronium hexafluorophosphate (see Chapter 3).
[b] For a tetrabranched MAP, omit *Cycle 3* and use 1 mmol each of Boc-amino acids, DCC, and HOBt.
[c] At this point the core matrix contains three levels of branched lysine and the eight amino groups will give an octabranched MAP.
[d] Do not cap the amino group if the peptide represents the N-terminal of a protein sequence.

4.2.4 Synthesis of MAPs by the indirect (modular) approach

In the indirect approach, the core matrix and the peptide antigens are prepared and purified separately (24). Both components (modules) are then combined to form the MAP. An advantage of this approach is that a free and purified peptide can be used. Another advantage is that the orientation of the peptide antigen can be arranged as in the native molecule such that there is a free and mobile C- or N-terminal.

The indirect approach (*Figure 3*) utilizes a purified peptide containing a cysteine residue at the N- or C-terminus to react with a core matrix containing chloroacetyl groups. Thiol alkylation of the chloroacetyl group at neutral to slightly alkaline pH combines both components to form a MAP construct containing the desired peptide epitope. The core matrix terminating with the chloroacetyl groups can be conveniently prepared by the solid-phase method using either Boc or Fmoc chemistry on a resin support (33–35). The chloroacetylated core matrix is cleaved by acid (35) from the resin support, and the crude product is usually sufficiently pure for conjugation to a peptide.

In a separate synthesis, the desired peptide epitope is prepared and purified to homogeneity (see Section 4.2.5). A synthetic peptide containing a cysteine at the N- or C-terminus is ideally suited for conjugation to the chloroacetyl groups of the core matrix (36, 37). Thiol alkylation to assemble both components is similar to the conventional conjugation of a peptide to a protein carrier. However, there are two major differences. First, unlike the protein carrier, the molecular weight of the chloroacetylated core matrix is low, less than 2000. Second, each addition of a peptide to the core matrix produces a relatively large increase in molecular weight. These two factors favour the completion of the reaction of the peptide with the core matrix and the resolution of the incomplete reaction products by gel permeation chromatography.

The indirect approach, using purified core matrix and peptide, is similar in concept and practice to the conventional two-step conjugation approach for ligating synthetic peptide to a protein carrier (25–31). Thus, many

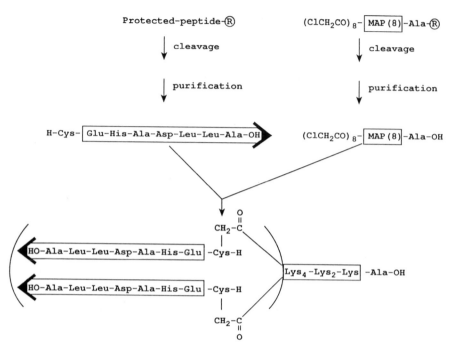

Figure 3. Synthesis of a MAP by the indirect, modular method. The chloroacetylated core matrix is combined with a purified peptide containing a cysteine at either the N- or C-terminal. The direction of the arrow indicates the orientation of the peptide from carboxy to amino terminal. The C-terminal 8-mer of human TGFα is used as an example. Note that the peptide immunogen mimics the parent protein with the C-terminal being free and flexible.

methodological variations of the modular approach can be envisioned. The method described for coupling a peptide containing a thiol moiety to a chloroacetylated core matrix (*Protocol 4*) is only one such variation. Extensions of this method include the use of another haloacetylated core matrix, for example a bromoacetylated or iodoacetylated core matrix. An iodoacetylated core matrix is known to be more reactive in the thiol alkylation reactions. However, chloroacetylated derivatives would be more stable than other haloacetylated derivatives in the peptide synthesis steps, such as the HF cleavage. Similarly, one can use other heterobifunctional cross-linking reagents such as MBS (28), *N*-succininimidyl 4-(*N*-maleimidomethyl)cyclohexane-1-carboxylate (29), succinimidyl 4-(*p*-maleimidophenyl)butyrate (30), and *N*-succinimidyl-3-(2-pyridyldithio)propionate (31). Whatever the reagent being used, the modular approach has the advantage of flexibility in producing a chemically defined peptide immunogen.

In *Protocol 4* the procedure for preparing a MAP by the indirect approach is given. This method gives yields of 50–80% and 20 to 40 mg of MAP for a 20-residue peptide.

Protocol 4. Synthesis of MAPs by the indirect (modular) method

Equipment and reagents

- HF apparatus (Peptide International, Peninsula Laboratories)
- Tetrabranched MAP core matrix: Boc-Lys$_2$-Lys-βAla-OCH$_2$-PAM resin (see *Protocol 3*; also available from Novabiochem, Bachem, and Applied Biosystems)
- Chloroacetic acid (Aldrich)

- DCC
- DCM
- DMF
- *N*-Methylpyrrolidone (NMP) (Aldrich)
- DIPEA
- Tributylphosphine (Aldrich)
- *t*-Butylmethyl ether

Method

1. Prepare the chloroacetic acid symmetric anhydride by adding DCC (0.12 mmol) to a solution of chloroacetic acid (0.24 mmol) in 5 ml of DCM. Filter off the dicyclohexyl urea after 1 h.

2. Add the chloroacetic anhydride solution to the deblocked and base-neutralized tetrabranched MAP core matrix (100 mg, 0.04 mmol) and agitate for 1 h.

3. Wash the resin with 10 ml amounts of DMF and DCM and dry it.

4. Cleave the chloroacetylated core matrix from the resin using neat HF (100%, no scavengers).[a] (**Caution**: see pp. 41–2.)

5. Extract the product in 10 ml of glacial acetic acid and precipitate it in cold *t*-butylmethyl ether. Repeat the precipitation twice and lyophilize.

6. Dissolve the chloroacetylated core matrix in 250 μl of NMP (4 μmol).[b]

7. Dissolve the thiol-containing peptide (8 μmol) in 1 ml of NMP.[c]

8. Add DIPEA (100 μmol) and tributylphosphine (50 μmol) to the peptide followed by the chloroacetylated core matrix (4 μmol based on chloroacetylated sites) and stir for 12 h.

9. Add 300 μl of water to clarify the solution and agitate it for an additional 6 h.

10. Add diethyl ether to precipitate the product.

11. Purify the MAP by gel permeation chromatography using 2–5% acetic acid as solvent.[d]

---

[a] Chloroacetylated core matrix is too soluble in organic solvents.
[b] All solvents should be deaerated and purged with nitrogen.
[c] The peptide is used in 2 M excess of the core matrix.
[d] Alternatively, repeat the precipitation and lyophilize. The incorporation of peptide can be determined by mass spectrophotometry.

4.2.5 Purification and analysis of peptides

Prior to immunization, removal of scavengers and by-products from the peptide must be performed. Coloured organic by-products, excess scavengers, and inorganic salts can be removed by organic solvent extraction, dialysis (for MAPs only), passage through a Sepak C-18 reverse-phase cartridge (Millipore), or desalting by ion-exchange resins. Gel permeation chromatography can be used to partially purify and desalt a synthetic peptide. A more effective method of purification is reverse-phase HPLC.

Peptide quality can be assessed by amino acid analysis, mass spectrometry, capillary electrophoresis, and ion-exchange HPLC. For production of anti-peptide antisera, highly purified peptide is desirable but not essential. Most practitioners have used crude or partially purified peptide mixtures with successful results. Nevertheless, a peptide with unambiguous characterization and purity will provide full confidence in the interpretation of the immuno-logical results.

4.3 The importance of peptide orientation

The N- and C-terminals of a protein are frequently the most flexible parts of the molecule and are known to be excellent sites for selection as peptide epitopes. The N-terminal of a protein is often variable within the same family of proteins making this segment suitable for the preparation of antibodies specific for a particular protein. On the other hand, the C-terminal peptide is often conserved and not processed in proteins (7), and such sequences may be useful for the preparation of antibodies that could cross-react with homolo-gous proteins of the same family. Because of the terminal charge (either NH_3^+ or CO_2^-) and the flexibility of these segments, the attachment of such peptides to the core matrix by either the N or C end could greatly affect the specificity of immune responses.

The importance of the correct orientation of a peptide attached to a protein carrier has been examined in peptide–protein conjugates (38). In general, the orientation of the peptide should mimic the parent protein. For peptides that are derived from the N-terminals or internal segments of protein molecules, the direct approach of linking the C-terminals to the core matrix would provide the best results. In such cases, the flexible end of a peptide would be at the N-terminal. For peptides that are C-terminals of proteins the direct approach will give an incorrect orientation because the free and flexible carboxyl end is attached to the core matrix. The indirect approach of prep-aration will overcome this deficiency; by positioning a cysteine residue at the N-terminal of a peptide, the peptide epitope can be attached to the chloro-acetylated lysine core matrix at its amino end to give the correct orientation. Thus, the indirect approach provides a solution to the deficiency of the direct approach when a C-terminal peptide is used.

The correct orientation has a direct relationship to the antigenicity of the peptide immunogen, as shown in the model system using an eight-residue peptide immunogen of the C-terminal of human transforming growth factor α (TGFα) (39). Conjugation to the core matrix at the N-terminus of the peptide would give more flexibility of the distal C-terminus (i.e. residues 46 to 50) than the proximal amino end linked directly to the core matrix. Such an orientation would also resemble the folded and native structure of TGFα Indeed, the anti-peptide antisera recognized both the peptide immunogen and TGFα, the parent protein (*Figure 4*). However, the anti-peptide antibody failed to recognize the 'reverse-immunogen' MAP even though the same sequence was present, but was attached to the core matrix in a reversed orientation via the C-terminal. The reason for such specificity was that the epitope for the antibody was located at the distal part of the immunogen containing the last four residues (residues 47–50) as well as the free carboxylic group.

4.4 The effect of the number of branches

The optimal number of branches of the MAP depends on several interacting factors. These include interchain steric hindrance, the length and conformation of the peptides, and the molecular size of the end product. Since lysine is asymmetrical, with a short α-amino arm and a long ε-amino arm, the

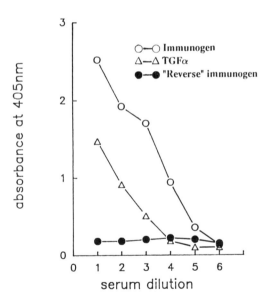

Figure 4. Specificity of antisera and the importance of peptide orientation. ELISA of antisera to a MAP derived from the C-terminal of human TGFα. Microplate wells were coated with: the immunogen (MAP-CEHADLLA), TGFα, and the 'reverse' immunogen (CEHADLLA-MAP). Serum dilutions were: 1, 1/16; 2, 1/64; 3, 1/256; 4, 1/1024; 5, 1/4096; 6, 1/16,384.

branching is also asymmetrical. The interchain steric hindrance due to the attachment of peptide can be examined by molecular models; no steric crowding is observed with branching up to eight dendritic arms. However, MAPs with 16 branches show steric hindrance. It has also been found that no steric crowding is observed during the synthesis of the peptides on a core matrix with eight branches. Difficulties such as incomplete coupling of each amino acid on the core matrix with 16 branches sometimes occur. More importantly, immune responses to MAPs with 16 branches have not shown superiority over those with fewer branches. Thus, MAPs with eight branches are preferred to those with 16 branches. Although steric hindrance is not a problem for the two-branch MAP, its immunogenicity is often inferior to that of the four- and eight-branch MAPs containing the same peptide immunogen. For this reason, two-branch MAPs are not used.

The choice of four or eight branches in the MAPs depends largely on the length and conformation adopted by the peptide immunogen. In the author's experience, the following suggestions can be made.

(a) For most peptides the length should be between 10 and 20 residues, and a MAP of four or eight branches is generally suitable. The desired epitope should be at the distal and flexible end of the peptide.

(b) For optimal results in peptides with 10 to 15 residues, MAPs with eight copies of the peptide are recommended. This recommendation is based largely on the consideration of the molecular size of the construct. The calculated molecular weight of a MAP containing eight copies of an 11-residue peptide would amount to about 10 500, which is similar to that of a small protein. However, for peptides with 16 to 20 residues, there is no overwhelming experimental evidence that favours MAPs with eight branches over those with four branches.

(c) For peptides with more than 21 residues, MAPs with four branches are recommended. The molecular size of a MAP containing four copies of a peptide antigen with 24 residues is about 11 000 Da. (Conformationally, the average number of amino acid residues that spans the length of a globular protein is 22–24.)

(d) The peptide should be relatively hydrophilic. Hydrophilic peptides are likely to be located on the surface of the protein and have desirable solubility. Also, note that soluble peptide always gives soluble MAPs.

(e) Although various lengths of peptide ranging from 5 to 37 amino acids have been used in the MAP approach with successful results, the most suitable length of peptide is in the range of 10 to 20 residues. The effective length appears to be dependent on the species of animal used for immunization. Outbred animals usually respond better than inbred animals, and rabbits are better responders than mice. Peptides of 15 to 20 residues provide the best results if mice are used for immunization, but peptides of 10 to 15 residues have been successful in outbred rabbits and guinea pigs. In general, longer peptides are preferred because part of the peptide often serves as a T-helper cell epitope to enhance immunogenicity. All the responses to the MAPs have been found to be T-cell-dependent, which indicates that the B-cell epitope also acts as a T-cell epitope. There may be greater success in the outbred animals because the repertoire of T-cell epitopes is large while the responses in inbred mice may be MHC-restricted and require specific T-helper epitopes.

4.5 Characteristics of immune responses to MAPs

4.5.1 Titres and specificity

The immunogenicity of MAPs is observable in the antibody titres. Specific antibody responses are generally produced in the immunization of outbred animals such as rabbits with a success rate of about 90% in the author's laboratory. Usually, the titres of end dilutions of 10^4 or higher are obtained after the second or third boostings. The antibodies obtained are usually specific in that they react with either the parent protein in solution or the denatured protein in immunoblots. The responses are usually lower in mice, particularly if they are inbred strains, due to the MHC restriction. If mice are used, it is advisable to immunize several strains.

The differences in immune responses in animals can be seen in the following examples. A MAP with eight copies of a 12-residue peptide consisting of three repeating tetrapeptides of Asn–Ala–Asn–Pro from the circumsporozoite (CS) protein of *Plasmodium falciparum* was found to be non-immunogenic in BALB/c mice but immunogenic in Black A/J mice and rabbits (11). Similarly, a MAP with four copies of an 11-residue peptide (RIQRGPGRAFV) from the gp120 protein of Human Immunodeficiency Virus-1 (IIIB strain) was found to be non-immunogenic in BALB/c mice but immunogenic to rabbits. However, immunogenicity in BALB/c mice was found with the extension of this 11-residue peptide to 17 residues (KSIRIQRGPGRAFVTIG). The response of the 17-residue peptide was due to the inclusion of the T-helper cell epitope.

A special point that needs to be stressed is the high monospecificity shown by the polyclonal and monoclonal antibodies elicited by peptides in the MAP approach. The monospecificity may be due to the homogeneity of the peptide inherent in MAP construction. Such monospecificity is particularly desirable for the identification and location of specific proteins in tissue sections by immunostaining. An example of antibody production for this purpose (14) is found in the localization and functional studies of the α subunit of the guanine nucleotide-binding protein G_0 (αO). Based on a partial sequence obtained from αO (AGISAKDV), a polyclonal antibody against this α subunit was raised using an octameric MAP. However, this particular segment also shares 56 to 66% of sequence homology with several other α subunits. Antiserum at a final dilution of 1:400 could detect αO in as little as 0.2 µg of G_0 and at a final dilution of 1:20 000 could detect the subunit in 1 mg of G_0 on immunoblots. Antiserum and affinity-purified antibody were specific to αO. Furthermore, no cross-reactivity was detected towards the α subunits of the stimulatory or inhibitory guanine nucleotide-binding regulatory proteins or transducin. The antiserum was then used to locate the distribution of αO in tissues and this allowed detailed immunocytochemical studies to show the tissue distributions and developmental appearance of αO.

4.5.2 Epitopes and monospecific antibodies

A question central to the preparation of peptide immunogens by the MAP approach is the understanding of the molecular basis for the immunogenicity of peptide attached to the core matrix, particularly the location of the dominant epitope of a peptide that binds to the antibodies induced by the MAP. Here the unique design of the MAP makes it possible to provide a clear answer to this question. We have mapped the epitopes of many antisera elicited by MAPs. A general conclusion based on MAPs consisting of peptides with 15 or fewer residues is that the dominant epitope is located on the flexible end, most distal from the core matrix. Furthermore, antigenicity decreases from the distal end to the proximal end, the least flexible part. Two examples are used below to support the generality of this conclusion.

In the first example (16), a 14-mer derived from the human T-cell receptor β-chain constant region was used in an octameric MAP and elicited highly immunogenic polyclonal and monoclonal antibodies in mice and rabbits. Nearly all of these antibodies reacted with the peptide in its monomeric as well as its octameric form. Moreover, these antibodies also reacted with the intact β-chain protein. To identify the epitope, a series of MAPs progressively shortened at the N-terminus was prepared (*Figure 5*). While 28 out of 28 monoclonal antibodies reacted with the full-length 14-residue MAP, six of the 28 reacted with the 11-residue MAP when the first three N-terminal residues were deleted, and one of the 28 antibodies reacted with the seven-residue MAP when the first seven N-terminal residues were deleted. Similarly, polyclonal antisera reacted less strongly with the 11-residue MAP and not at all with the seven-residue MAP. No antibodies were found to react with the core matrix. The combined results demonstrate that the amino acid residues representing the distal end of the 14-residue octameric MAP, which are most exposed and flexible, constitute the epitope recognized by the antibodies in the polyclonal sera and by the monoclonal antibodies.

In the second example (22), the antigenic site was determined by the Pepscan approach, in which the entire antigenic sequence is 'windowed' with the use of a series of overlapping synthetic peptides differing by one amino acid (see Chapter 7). A 20-residue peptide of the neutralizing epitope of the coat protein, VP1 (141–160), of the FMDV was incorporated in an octameric MAP. The MAP was compared with conventional peptide–carrier conjugates

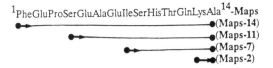

^{1}PheGluProSerGluAlaGluIleSerHisThrGlnLysAla^{14}-Maps

		Recognition by (ELISA)		
Peptide	Size (aa)	Mab	Pab	Control
Maps	14	22/22	+	-
	11	4/22	+/-	-
	7	1/22	-	-
	2	0/22	-	-
Monomeric	14	15/22	+	-
	11	6/22	+/-	-
Polylysine	14	1/22	-	-

Figure 5. Epitopes of polyclonal (Pab) and monoclonal (Mab) antibodies to a 14-mer from the human T-cell receptor β chain. The MAPs used in the ELISA contained 14, 11, 7, and 2 residues and the polylysine peptide was the monomeric peptide coupled to polylysine with a water-soluble carbodiimide. The negative control is a mean of 12 values obtained from six different monoclonal antibodies with irrelevant specificity.

in which VP1 (141–160)-cysteine was coupled by MBS or by gluteraldehyde to KLH. Two rabbits were inoculated with each immunogen. The MAPs induced a very high anti-peptide response 8 weeks after immunization (titre of 10^4), while the peptides conjugated with KLH induced a lower response. However, neutralization titres of all antisera were similar. The epitopes for these antisera were then determined by the Pepscan method using overlapping eight-residue peptides from the region 134–165 of VP1 of FMDV type O (*Figure 6*). The results revealed that a large population of antibodies raised against the MAP were specific for the flexible N-terminal amino acids and that both rabbits gave a similar response (*Figure 6*a, b). In contrast, antibodies raised against the peptides conjugated with KLH reacted with different sequences, and each rabbit gave a different response (*Figure 6*c, d, e, and f). Thus, two conclusions can be drawn from this example. First, these results confirm those of the previous example that the distal and flexible end of the MAP is the most antigenic. In this example, peptides ranging from eight to 20 residues consistently show that the distal end is the epitope. Second, MAPs produce a better controlled anti-peptide response than that of the conventional peptide–protein carrier conjugates. This result is not unexpected since a MAP is small, chemically defined, and homogeneous in contrast to the conventional peptide–protein conjugate, which is likely to be large, chemically ambiguous, and heterogeneous. Thus, the better controlled antipeptide response resulting from the MAP is particularly suited for the production of monospecific antibodies for biochemical studies.

In summary, the combined results suggest that for short peptides (15 residues or fewer) the distal and most flexible portion, away from the site of

a,b   VPNLRGDLQVLAQKVARTLP-[MAP]
c,d   [KLH/MBS]-CVPNLRGDLQVLAQKVARTLP
e,f   [KLH/GDA]-CVPNLRGDLQVLAQKVARTLP

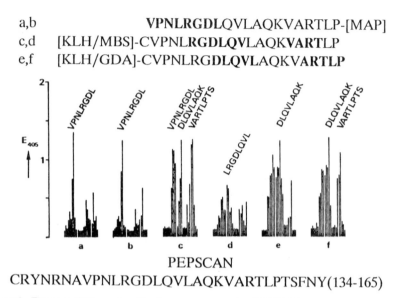

PEPSCAN
CRYNRNAVPNLRGDLQVLAQKVARTLPTSFNY(134-165)

Figure 6. The specificity and antibody responses to the FMDV VP1 141–160 peptide using the Pepscan method. Each line represents the ELISA value of an 8-mer in a set of overlapping peptides starting at residue 134. Rabbits a and b were immunized with the MAP while the other rabbits were immunized with peptide conjugated to KLH (with MBS in the case of c and d, and with glutaraldehyde in the case of e and f). The major epitope sequences are given.

attachment to the core matrix, is most likely to be the epitope. This conclusion may not be applicable to longer peptides (more than 15 residues) which may be folded into more defined structures with the distal ends looped to the C-terminals. Nevertheless, a direct and useful strategy to prepare short peptide immunogens can be derived from these results: that is, using the MAP approach the desired epitope of a short peptide should be placed on the distal end of the MAP construct.

4.6 MAP applications to vaccines

4.6.1 Requirements for B- and T-cell epitopes

A mechanism to generate protection against infection is the production of neutralizing antibodies. However, the peptide antigen that carries only the neutralizing epitope (the B-cell epitope) may not be sufficiently immunogenic. As discussed earlier, the B-cell epitope is often conjugated to a protein carrier to enhance its immunogenicity. The carrier thus serves a T-helper cell function that assists the proliferation of the plasma cells and the secretion of antibodies of the desired specificity. Since the dominant T-helper cell epitope can be identified, a major goal in the development of synthetic peptide vaccines using the MAP approach is the design of suitable models for incorporating the neutralizing B-cell epitope and the dominant T-helper cell epitope in an optimal arrangement, orientation, and stoichiometry. In the following example, several MAP models containing both B- and T-cell epitopes are used to illustrate the versatility of the oligomeric lysine core matrix for this purpose.

4.6.2 MAPs and malaria

Each year 200 million people are afflicted with malaria, with a fatality rate of 1–2%. Malaria is stage-specific. The first stage of the malaria parasitic cycle in humans involves the sporozoite which is injected by the mosquito into the host's bloodstream. A vaccine against the sporozoite stage appears to be most advantageous in blocking infection in humans and in preventing transmission of the disease. However, the development of such a vaccine is limited by several technical factors. These factors include the difficulty in obtaining large amounts of the surface antigens of the sporozoites and, since malaria is species-specific, it has been difficult to culture sporozoites that specifically infect humans. Thus, a synthetic vaccine based on peptides appears to be necessary and also quite attractive, since the technology is relatively simple and could easily be transferred to the countries most affected by malaria.

An important requirement for synthetic peptide vaccine development is the identification of the appropriate B- and T-cell epitopes so that they may be attached to the MAP for experimental verification of vaccine efficacy in a well defined animal model for a specific infectious disease. These requirements for vaccine development can be met in malaria, a disease for which no practical immunoprophylaxis has yet been developed. Previously, it was shown that protective immunity against rodent, simian, and human malaria sporozoites can be induced by immunization with irradiated sporozoites (40). The major surface protein of sporozoites is the CS protein, and the antibodies directed against the CS protein neutralize the infectivity of the parasites and inhibit their entry into hepatocytes (41). Thus, the CS protein has become a logical choice for the development of vaccines against the sporozoite stage of malaria (42, 43). The B-cell epitope of the CS protein of the rodent malaria parasite, *Plasmodium berghei*, is contained within the repeating domain (44) of the CS protein, a feature common to CS proteins of all malaria species (45, 46). Mice immunized with a synthetic vaccine consisting of the B-cell epitope attached to tetanus toxoid as a carrier have developed high antibody titres against the native CS protein and show resistance to challenge with 10^3 sporozoites (47). However, vaccination attempts in humans, using a similar approach, have failed to induce good antibody titres. This result points to the limitation of using a protein carrier (tetanus toxoid) that leads to epitopic suppression caused by the strong immune response to the carrier protein (48).

Recently, several T-helper cell epitopes of the CS protein of *P. berghei* have been identified (49), thus providing the opportunity to incorporate B- and T-cell epitopes in MAPs and test their efficacy in the rodent malaria model. MAPs containing T- and B-cell epitopes derived from the CS protein have been made. The results obtained were compared with those obtained by immunization with the monomeric peptide (containing only one copy of the B- or T-cell epitopes) and with earlier results obtained by immunization with the recombinant CS protein and irradiated sporozoites of this plasmodial species.

The immunodominant B-cell epitope of the CS protein of *P. berghei* can be represented by a tandemly repeating peptide of 16 residues while a peptide that contains a T-helper cell epitope recognized by mice of four different H-2 haplotypes is contained between residues 265 and 276 of the CS protein (*Figure 7*). These T- and B-cell epitopes were incorporated into 10 MAPs in several arrangements and stoichiometric ratios. These included two MAPs containing four or eight tandemly connecting B- and T-cell epitopes (models BT(4) and BT(8)) and two others containing similar epitopes but with the orientation of the B- and T-cell epitopes reversed (models TB(4) and TB(8)). Two MAPs contained a different stoichiometry of the B- and T-cell epitopes; B(8)T contained eight copies of the B-cell epitope and only one copy of the T-cell epitope, while model T(8)B contained eight copies of the T-cell epitope and only one copy of the B-cell epitope (*Figure 7*). As controls, MAPs containing either B- or T-cell epitopes alone in either the tetrameric or octameric arrangement were used (49).

Earlier experiments had shown that immunization with the *P. berghei* B-cell epitope alone, either as a monomer (i.e. the 16-residue peptide) or as an octameric MAP, does not elicit antibody responses in several inbred strains of mice bearing different H-2a haplotypes. This genetic restriction was over-come by the addition of a T-helper cell epitope to the B-cell epitope. For example, A/J mice (H-2a) did not respond to the octameric MAP containing the B-cell epitope, but A/J mice that had been primed with a single dose of *P. berghei* sporozoites produced a very strong secondary response when boosted with a synthetic peptide containing in tandem the B-cell epitope and the T-cell epitope. Thus, only the di-epitope MAPs containing both T- and B-cell epitopes were expected to produce immunological responses in the A/J mice.

Groups of five mice of the A/J strain were each immunized with one of the 10 MAPs. Controls were immunized with the BT-monomer, or with a mixture consisting of an equimolar ratio of B(8) and T(8). The sera obtained from each group of animals were pooled and assayed for their reactivity with the recombinant CS protein by immunoradiometric assay and with gluteraldehyde-fixed sporozoites by indirect immunofluorescence assay.

Based on the antibody responses 21 days following the first antigen dose, the MAPs could be classified into two groups according to their immuno-genicity (*Figure 8*). The MAPs which contained equimolar amounts of T- and B-cell epitopes, linked in tandem, were highly immunogenic and elicited high antibody titres. The best immunogen, BT(4), produced serum antibody levels detectable at dilutions greater than 10^5 and the others induced serum titres

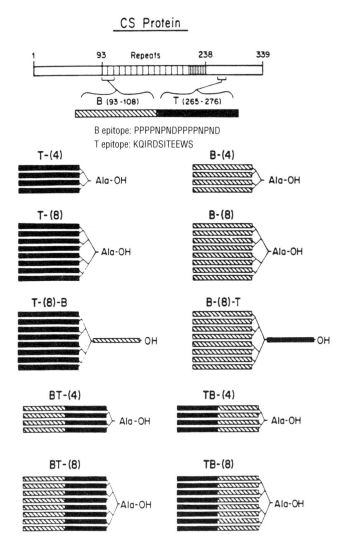

Figure 7. The structure of *P. berghei* CS protein showing B- and T-cell epitopes and schematic representations of 10 MAPs. For convenience each MAP is named according to the type and number of epitopes it contains, e.g. T-(8) has eight copies of the T-cell epitope.

between 6.4×10^4 and 2.5×10^4. The poorly immunogenic MAPs included the two in which there were different ratios of B- and T-cell epitopes, B(8)T and T(8)B. MAPs B(4) and B(8) were non-immunogenic as expected because they did not contain a T-helper cell epitope to overcome the genetic restriction of the immune response in A/J mice. There was a very poor antibody response to T(4), T(8), and to TB-monomers. Thus, a covalent high molecular weight

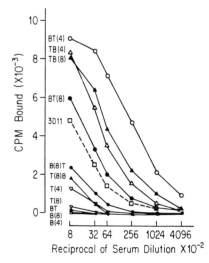

Figure 8. Secondary responses of groups of mice immunized with different MAPs. Groups of five mice of H-2a haplotype were injected with 50 μg of each MAP or a mixture of MAPs T-(8) and B-(8) emulsified in complete Freund's adjuvant on day 0 and boosted with 50 μg of the same MAP in incomplete Freund's adjuvant on day 21. Sera collected 21 days later were pooled and the antibody titres determined using an immunoradiometric assay with recombinant CS protein as antigen. For comparison, the titres of antisera to the sporozoite 3D11 are shown.

structure containing multiple copies of both B- and T-cell epitopes such as BT(4) or TB(4) was required for immunogenicity since the mixture of equimolar amounts of B(8) and T(8) failed to elicit any antibody responses (49).

In general, the antibody responses observed 21 and 34 days after a second dose of the MAPs were significantly higher, but the ranking order of immunogenicity was similar to that observed in the primary responses. The most immunogenic MAP, BT(4), produced antibodies that reacted at serum dilutions of 5×10^5, while the other three di-epitope MAPs, containing equimolar amounts of T- and B-cell epitopes, produced titres between 10^5 and 4×10^5. Animals immunized with the mono-epitope MAPs such as B(4) and B(8) MAPs, remained non-responders. The immunofluorescence titres of the sera were measured at 34 days after the booster injection. The BT(4)-injected mice had titres of 128×10^3, which are much higher than the $(5–15) \times 10^3$ titres usually found in the serum of mice hyperimmunized with irradiated sporozoites. Immunization of A/J mice with similar doses of recombinant CS protein, encompassing amino acids 81–277, resulted in much lower titres (2×10^3).

The apparent greater immunogenicity of BT(4) than that of other di-epitope models appears to be consistent with the general conclusion derived from the previous discussion. That is, BT(4) contains the favourable B-cell epitope at the distal end of the MAP construct and in the less crowded tetrameric format (BT has 28 residues). BT(4) produced antibody titres about 50-fold higher than the bulkier BT(8). On the other hand, there is little difference in response between the reversely oriented models TB(8) and BT(8). The B-cell epitope of the CS protein is exceptionally rich in proline (50%), while the T-cell epitope has a strong propensity for amphipathic helix formation. Although the results indicate that there is no advantage in increasing the number of MAP branches from four to eight, the effects on the immune response of the number and orientation of the B- and T-cell epitopes in the MAP models may be sequence-dependent and may require optimization in a given experimental system.

The efficacy of the MAPs as a vaccine was determined by an intravenous challenge of mice with 2000 *P. berghei* sporozoites 35 days after the secondary immunization with the MAPs. In the group of mice immunized with BT(4), 80% of the mice were protected, while in the group of mice immunized with the other three di-epitope MAPs, there was 50 to 60% protection. No protection was observed in the mice immunized with the mono-epitope MAPs. The anti-sporozoite antibody levels, as determined by immunofluorescence assays before the challenge, correlated well with the degree of protection.

Comparison of protein carrier versus MAP vaccines is seen in studies of *Plasmodium falciparum* synthetic vaccines. Human clinical trials have been completed using a synthetic vaccine consisting of $(AsnAlaAsnPro)_3$, which is the B-cell epitope of the CS protein of *P. falciparum* that has been coupled to the tetanus toxoid carrier (42). The frequency and magnitude of the antibody response to the peptide increased with the dose of antigen; however, due to the toxicity of tetanus toxoid, the vaccine dose could not be increased. One of the disappointing findings of this trial was that the antibody levels were much lower than those in mice given the same vaccine. This may be due in part to the relatively high dose of the conjugate given to mice, or because the human volunteers had been previously vaccinated with tetanus toxoid, and their response to the B-cell epitope was therefore inhibited by antibodies to the tetanus toxoid. The di-epitope MAPs should overcome both these problems of epitopic suppression and carrier toxicity. Furthermore, the MAPs are chemically defined, contain a high density of epitopes, and, as shown here, can produce high levels of parasite-specific protective antibodies.

5. Immunization procedures

The choice of the animal is often dictated by the quantity of antiserum required. Twenty millilitres of serum can be drawn from rabbits (e.g. New Zealand red or white rabbits) and 5 ml from guinea pigs without harmful effects. Inbred and outbred mice are less expensive to maintain but only 0.5 ml of serum can be obtained.

For eliciting both poly- and monoclonal anti-peptide antibodies, the immunogen is emulsified with an adjuvant and a vehicle and administered intramuscularly, intradermally, subcutaneously, or intraperitoneally into the animal. Polyclonal antisera have advantages over monoclonal antibodies as they require less effort to make and are more reliable sources of high-titre and high-affinity antibody, but monoclonals have the advantages of permanency and high specificity. Both types are suitable as reagents for various assays such as immunoblotting.

Typically, the specific antibody response appears one week after the primary immunization. In the case of peptide immunogens, the titres are of low-affinity IgGs and these will level off and decrease at week two. These antibodies are generally not suitable as reagents. To achieve high-titre IgGs, a booster at week four is necessary. B-cell populations in the course of immunization produce progressively higher-affinity antibodies by somatic hypermutation and selection and some B-cells are differentiated into memory cells. The booster injections activate memory B-cells and cause the B-cells to proliferate and to switch to IgG antibody. One or more additional boosters can be administered at two to four week intervals after the initial priming and boosting injections if high-titre antiserum is not obtained. The animal is bled and serum is prepared prior to and following the primary immunization, and following each booster injection.

5.1 Production of antibodies

Adjuvants greatly enhance the specific antibody titre as they allow the antigen to be released slowly, thus ensuring the prolonged presence of antigen to stimulate the immune system. Freund's adjuvant is used in the preparation of the immunogen because it induces a long-lasting antibody titre that is often still measurable several weeks or more after boosting. The presence of mycobacteria in complete Freund's adjuvant activates the T-cell population providing the necessary lymphokines for B-cell stimulation and maturation. Incomplete Freund's adjuvant is used for booster injections. Examples of immunization procedures are given below.

Mice. Ten- to twelve-week-old mice are injected intraperitoneally with 10 μg of an antigen in 100 μl of PBS emulsified with an equal volume of complete Freund's adjuvant. Four weeks later, the mice are boosted intraperitoneally with 10 μg of the antigen in 100 μl of PBS emulsified with an equal volume of incomplete Freund's adjuvant. One week later, all immunized mice are bled from the tail vein and their sera collected for immunological testing. The mouse with the highest serum titre can be selected for the production of monoclonal antibodies. (Three days prior to fusion one more booster injection is given intravenously or intraperitoneally with 10 μg of the antigen in PBS.)

Guinea pigs. Female Hra/Dunkin Hartley guinea pigs, 400–450 g, are injected subcutaneously and intradermally at multiple sites with 100 μg of an antigen in 200 μl of PBS emulsified with an equal volume of complete Freund's adjuvant and boosted with the same dose emulsified with incomplete Freund's adjuvant four weeks after the initial immunization and at three-week intervals thereafter. Sera are collected two weeks after each boost.

Rabbits. Female rabbits are injected subcutaneously at multiple sites along the back with 0.5 g of antigen in 0.5 ml of PBS emulsified in an equal volume of complete Freund's adjuvant and are boosted at four-week intervals with the same dose in incomplete Fruend's adjuvant. Sera are collected from the marginal ear vein two weeks after each boost.

The production of monoclonal antibodies requires the identification of hybridoma cell clones secreting a single type of antibody molecule directed against the antigen of interest. Propagation of these clones is performed in tissue culture and the medium containing the secreted antibodies can be harvested. This produces low amounts of antibody. For large quantities, it is best to propagate the cloned cells in mice which then produce ascitic fluid containing high concentration of the antibody (50, 51)

A poor immunological response after several booster injections may be overcome by: increasing the amount of antigen; changing adjuvant or vehicle; increasing the number of animals immunized; changing the host, e.g. from mice to guinea pigs or rabbits. However, in the majority of cases, a poor response is due to the peptide being a poor immunogen. Choosing another sequence or using a different approach for preparing the antigen–carrier complex may resolve this difficulty.

5.2 Preparation and storage of serum and ascitic fluid

Serum is prepared by allowing the blood to stand for 3–5 h at room temperature followed by overnight cold storage (4 °C) to form a clot. The clot is then gently removed with the aid of a wooden applicator stick, and the serum is clarified by centrifugation for 10 min at 2500 g at 4 °C. The supernatant serum is removed and assayed for activity.

Multiple harvests of *ascitic fluid* are pooled, clarified by centrifugation (10 min at 1500 g, at room temperature), and heat-inactivated for 45 min in a 56 °C water-bath. Clots are removed and the solution is clarified again.

Antibodies, in a buffered condition, are relatively stable at −20 °C, and can generally be stored for long periods of time, particularly in serum, plasma, and ascitic fluid. Low concentrations of NaN_3 (0.015–0.10 M final concentration) are often added to these solutions to prevent growth of microorganisms. However, NaN_3 is poisonous and one must consider the future use of the antibody as NaN_3 is known to interfere with many biological assays and experiments. To sterilize solutions containing antibodies, it is best to filter sterilize (0.45 μm pore size). Aliquot and freeze the sample at −70 °C. For frequent use, antiserum, or ascitic fluid containing NaN_3 is aliquoted in smaller volumes and stored at 4 °C for up to 4–6 months. Repeated freeze–thaw routines are to be avoided.

The IgG fraction can be obtained by salt precipitation and anion exchange chromatography or Protein A chromatography, and specific antibodies can be purified by immunoaffinity methods (see Chapter 5, Section 5).

References

1. Ada, G. L. (1990). *The Lancet*, **335**, 523.
2. Brown, F. (1990). *The Lancet*, **335**, 587.
3. Milich, D. R. (1989). *Adv. Immunol.*, **45**, 195.
4. Sela, M. (1969). *Science*, **166**, 1465.
5. Arnon, R., Maron, E., Sela, M., and Anfinsen, C. B. (1971). *Proc. Natl. Acad. Sci. USA*, **68**, 1450.
6. Sutcliffe, I. G., Shinnick T. M., Green, N., and Lerner, R. (1983). *Science*, **219**, 660.
7. Walter, G., Scheidtman, K., Carbone, A., Laudano, A. P., and Doolittle, R. F. (1980). *Proc. Natl. Acad. Sci. USA*, **77**, 5197
8. DiMarchi, R., Brooke, G., Gale, C., Cracknell, V., Doel, T., and Mowat, N. (1986). *Science*, **232**, 639.
9. Jacob, C. O., Arnon, R., and Sela, M. (1985). *Mol. Immunol.*, **22**, 1333.
10. Nussenzweig, V. and Nussenzweig, R. S. (1989). *Adv. Immunol.*, **45**, 283.
11. Tam, J. P. (1988). *Proc. Natl. Acad. Sci. USA*, **85**, 5409.
12. Tam, J. P. (1989). In *Methods in enzymology* (ed. P. M. Conn), Vol. 168, pp. 7–14. Academic Press, London.
13. Francis, M. J., Hastings, G. Z., Brown, F., McDermed, J., Lu, Y. A., and Tam, J. P. (1991). *Immunol.*, **73**, 249.
14. Chang, K. J., Pugh, W., Blanchard, S. G., McDermed, J., and Tam, J. P. (1988). *Proc. Natl. Acad. Sci. USA*, **86**, 4929.
15. Zavala, F., Romero, P. J., Ley, V., Nussenzweig, R., Nussenzweig, V., Tam, J. P., and Barr, P. J. (1988). In *Vaccines 6*, p. 66. Cold Spring Harbor Laboratories, NY.
16. Posnett, D., McGrath, H., and Tam, J. P. (1988). *J. Biol. Chem.*, **263**, 1719.
17. Tam, J. P. and Zavala, F. (1989). *J. Immunol. Methods*, **124**, 53.
18. Tam, I. P. (1989). In *Synthetic peptides: approaches to biological problems*, (ed. J. P. Tam and E. T. Kaiser), pp. 3–18. Alan R. Liss, New York.
19. Tam, J. P. (1989). In *Vaccines 1*, pp. 21–5. Cold Spring Harbor Laboratories, NY.
20. Tam, J. P., Clavijo, P., Lu, Y. A., Nussenzweig, R. S., Nussenzweig, V., and Zavala, F. (1990). *J. Exp. Med.*, **171**, 299.
21. Tam, J. P. and Lu, Y. A. (1989). *Proc. Natl. Acad. Sci. USA*, **86**, 9084.
22. Schaaper, W. M. M., Lu, Y. A., Tam, J. P., and Meloen, R. H. (1990). In *Peptides: chemistry, structure and biology* (ed. I. E. Rivier and G. R. Marshall), pp. 765–6. ESCOM Science Publishers, Leiden.
23. Tam, J. P. and Lu, Y. A. (1990). In *Innovation and perspectives in solid phase peptide synthesis—peptides, polypeptides, and oligonucleotides: macro-organic reagents and catalysts* (ed. R. Epton), pp. 351–70. SPCC (UK), Birmingham.

24. Lu, Y. A., Clavijo, P., Galantino, M., Shen, Z. Y., Liu, W., and Tam, J. P. (1991). *Mol. Immunol.*, **28**, 623.
25. Avrameas, S. and Ternyck, T. (1969). *Immunochemistry*, **6**, 53.
26. Bernatowicz, M. S. and Matsueda, G. R. (1986). *Anal. Biochem.*, **155**, 95.
27. Lee, A., Powell, J. E., Tregar, G. W., Niall, H., and Stevens, V. (1980). *Mol. Immunol.*, **17**, 749.
28. Kitagawa, T. and Aikawa, T. (1976). *J. Biochem.*, **79**, 233.
29. Yoshitake, S., Yamada, Y., Ishikawa, E., and Masseyeff, R. (1979). *Eur. J. Biochem.*, **101**, 395.
30. Martin, F. and Papahadjopoulos, D. (1982). *J. Biol. Chem.*, **257**, 286.
31. Carlsson, J., Drevin, H., and Axen, R. (1978). *Biochem. J.*, **173**, 723.
32. Beinfield, M. C. and O'Donohue, T. L. (1989). In *Methods of enzymology* (ed. P. M. Conn), Vol. 168, pp. 3–7. Academic Press, London.
33. Merrifield, R. B. (1986). *Science*, **232**, 341.
34. Merrifield, R. B. (1963). *J. Am. Chem. Soc.*, **85**, 2149.
35. Tam, J. P., Heath, W. F., and Merrifield, R. B. (1983). *J. Am. Chem. Soc.*, **105**, 6442.
36. Linder, W. and Robey, F. A. (1987). *Int. J. Pept. Prot. Res.*, **30**, 794.
37. Robey, F. A. and Fields, R. L. (1989). *Anal. Biochem.*, **177**, 373.
38. Dryberg, T. and Oldstone, M. B. (1986). *J. Exp. Med.*, **164**, 1344.
39. Tam, J. P. (1987). *Int. J. Pept. Res.*, **29**, 421.
40. Yoshida, N., Nussenzweig, R. S., Potocnjak, P., Nussenzweig, V., and Aikawa, M. (1980). *Science*, **207**, 71.
41. Egan, J. E., Weber, J. L., Ballou, W. R., Hollingdale, M. R., and Majarian, W. R. (1987). *Science*, **236**, 453.
42. Zavala, F., Tam, J. P., Hollingdale, M. R., Cochrane, A. H., Quakyi, I., Nussenzweig, R. S., and Nussenzweig, V. (1985). *Science*, **228**, 1436.
43. Ballou, W. R., Hoffman, S. L., Sherwood, J. A., Hollingdale, M. R., Neva, F. A., Hockmeyer, W. T., Gordon, D. M., Schneider, I., Wirtz, R. A., Young, J. F., Wasserman, G. F., Reeve, P., Diggs, C. L., and Chulay, I. D. (1987). *The Lancet*, **i**, 1277.
44. Eichinger, D. L., Arnot, D. E., Tam, J. P., Nussenzweig, V., and Enea, V. (1986). *Mol. Cell. Biol.*, **6**, 3965.
45. Weber, J. L., Egan, J. E., Lyon, J. A., Wirtz, R. A., Charoenvit, Y., Maloy, W. L., and Hockmeyer, W. T. (1987). *Exp. Parasitol.*, **63**, 295.
46. Dame, J. B., Williams, J. L., McCutchan, T. F., Weber, J. L., Wirtz, R. A., Hockmeyer, W. T., Maloy, W. L., Haynes, J. D., Schneider, I., Roberts, D., Sanders, G. S., Reddy, E. P., Diggs, C. L., and Miller, L. H. (1984). *Science*, **225**, 593.
47. Zavala, F., Tam, J. P., Barr, P. J., Romero, P. J., Ley, V., Nussenzweig, R. S., and Nussenzweig, V. (1987). *J. Exp. Med.*, **166**, 1591.
48. Herzenberg, L. A. and Tokuhisa, T. (1983). *J. Exp. Med.*, **155**, 1730.
49. Romero, P. J., Tam, J. P., Schlesinger, D., Clavijo, P., Barr, P. J., Nussenzweig, R. S., Nussenzweig, V., and Zavala, F. (1988). *Eur. J. Immunol.*, **18**, 1951.
50. Liddell, J. E. and Cryer, A. (1991). *A practical guide to monoclonal antibodies*. Wiley, Chichester.
51. Peters, J. H. and Baumgarten, H. (1992). *Monoclonal antibodies*. Springer-Verlag, New York.

Preparative immunoaffinity techniques

MICHAEL R. PRICE and KAMI BEYZAVI

1. Introduction

Immunoaffinity chromatography is one of the most powerful separative procedures for the purification of antigens or antibodies. The technique exploits the exquisite immunological specificity of their recognition and binding, which occurs even in complex mixtures. The procedure is highly empirical, but often separations resulting in purification of a product by several thousand-fold may be achieved by using simple chromatographic techniques. The technique has found application in the purification of diverse macromolecules. This has been particularly enhanced over the last decade or so, following the development of hybridoma technology for the production of stable cell hybrids secreting homogeneous monoclonal antibodies of identical immunological and biochemical characteristics (1). Moreover, it is feasible to identify the epitopes for many monoclonal antibodies and, in the case of protein antigens, to prepare synthetic peptides corresponding to the native epitopes. This is being exploited in an attempt to enhance our understanding of the antigen–antibody interaction, and at the practical level to provide novel approaches in immunoaffinity techniques.

1.1 General principles of immunoaffinity chromatography

Immunoaffinity chromatography is a highly specialized type of adsorption chromatography (2). In this method one or more of the components in a mixture to be separated interacts with the insoluble particles (the matrix) comprising the chromatographic medium, which is usually packed into a chromatographic column. Unadsorbed components in the mixture remain in the mobile liquid phase which then can be easily separated from the matrix. In one form of immunoaffinity chromatography an antibody is immobilized on to the insoluble chromatographic matrix. The corresponding soluble antigen in the mixture to be resolved can be specifically adsorbed to the substituted matrix following immunological recognition and binding, and the non-bound moieties (the contaminants) are then simply washed away. The complex between the insoluble immunoadsorbent and antigen is subsequently dissociated and the purified antigen obtained. Alternatively, antigen (rather than antibody) may be immobilized on the matrix and this is used to adsorb and separate its antibody from a complex solution (e.g. an antiserum), so that highly purified antibody may then be eluted.

2. Choice of gels and conjugation procedures

There are a number of options in the design of any particular purification protocol. However, attention here will be focused upon a number of examples presented to illustrate the strategies adopted and technologies employed for small-scale separations required in a research laboratory.

2.1 Theoretical requirements

The ideal gel matrix should have the following general properties:

(a) chemical and mechanical stability—the gel should be insoluble over a range of conditions of pH, detergents, varying ionic strengths and solvent concentrations

(b) rigidity of matrix to withstand high flow rates and differing hydrodynamic pressures

(c) presence of functional groups which can be reacted under non-denaturing conditions with antibody (or antigen) resulting in stable covalent linkages

(d) high and uniform porosity to ensure large effective surface area and accessibility of macromolecules within the matrix

(e) resistance to microbial and enzymatic degradation

It is interesting to note that some of the bizarre geometrical constructs visualized by the originators of chaos theories (e.g. a Menger sponge—a cubic solid-looking lattice with infinite surface area but zero volume! (3)) would in fact make excellent chromatographic matrices for affinity separations (with high porosity, rigidity, surface area, etc.) although it is beyond the scope of this chapter to follow this further.

Commercially available matrices can be divided into two groups: natural and synthetic supports.

2.2 Natural matrices

These include agarose, cellulose, glass, silica, and alumina (*Table 1*) (4, 5). The polysaccharide, agarose, which has been widely used for many years, has a primary structure consisting of alternating residues of D-galactose and 3-anhydrogalactose. The matrix contains primary and secondary alcohols which can be used for activation and attachment of ligands. Because of its

Table 1. Some commercially available matrices

	Trade name	Company
Natural matrices		
Agarose	Sepharose	Pharmacia
Cellulose	Cellulofine	Amicon
Controlled pore glass	CPG	CPG
Surface-modified controlled pore glass	PROSEP	Bioprocessing Ltd
Synthetic matrices		
Polyacrylamide	Bio-Gel P	Bio-Rad Laboratories
Poly 'tris acrylamide'	Trisacryl	IBF Biotechnics
Dextran/acrylamide	Sephacryl	Pharmacia
Vinyl polymer	Toyopearl	TosoHaas Corporation
Polystyrene	Poros	PerSpective Biosystems
Microporous membranes	MAC	Amicon

enhanced structural stability, cross-linked agarose is usually preferred over the non-cross-linked variety for most affinity preparations although cross-linking results in about 30 to 50% loss of the hydroxyl groups which are available for activation.

Some of the disadvantages of agarose matrices for use in affinity purification processes, which are particularly evident on scale-up, include mechanical instability, susceptibility to microbial attack, and variations in pore size which can lead to trapping of macromolecules.

Inorganic rigid matrices such as controlled pore glass (CPG) are more suitable to large-scale affinity preparations. CPG consists of porous granules of high-silica glass permeated by interconnecting pores of uniform and precisely controlled size. These matrices are chemically and mechanically stable (and virtually incompressible), and are insoluble over a wide range of conditions. Pore diameters can be selected within a range 70 Å to 3000 Å (with 80% of the pores showing a deviation of less than ±10% from the nominal diameter). For the covalent attachment of ligands, the surface of CPG (Si–OH) can be modified by reacting the silanol groups of the glass with suitably functionalized silanes.

2.3 Synthetic matrices

Synthetic matrices are prepared by polymerization of functional monomers to give products suitable for affinity purification processes. In some cases co-polymer derivatives have been produced from the combination of a naturally occurring matrix such as agarose and synthetic monomers. *Table 1* summarizes some of the available synthetic supports.

Synthetic matrices have good physical and chemical stability and can tolerate high pressures. Many can withstand organic solvents and extremes of pH without matrix decomposition. Because of their synthetic nature, the base matrices are usually microbe resistant. The incorporation of monomers with suitable functional groups can provide activation sites on these supports for ligand immobilization. Monomers that contain primary or secondary hydroxyl groups are normally used in the production of synthetic supports.

2.4 Activation chemistry

Activation chemistry is defined as a process to modify a matrix chemically so that a ligand can be bound covalently to the matrix (2, 6). The activation chemistry must be compatible with both the matrix and the ligand. The following must be considered in the selection of an activation chemistry:

(a) It must not have any adverse effects upon the biological activity of the ligand.

(b) It must result in a stable bond between the matrix and the ligand—unstable bonding will cause leakage of the ligand from the matrix into the final product and reduce the useful life of the affinity support.

(c) It must not induce non-specific binding of contaminating molecules. Non-specific binding is either produced by hydrophobic or ionic interactions; therefore the activation must produce a matrix which is hydrophilic and charge free.

(d) It must bind the ligand efficiently even at low concentration. This is particularly important when the ligand is scarce and expensive.

(e) It must not alter the physical properties of the matrix.

(f) The reagents used in the activation chemistry should, if possible, be non-hazardous and readily available. The process should be easily scalable and fast.

Table 2 summarizes some of the more commonly used activation chemistries employed for immobilizing ligands to a matrix and many commercially available matrices may be obtained in a pre-activated state (e.g cyanogen-bromide-activated Sepharose 4B from Pharmacia).

Table 2. Coupling chemistries for linking ligands to matrices

Activation chemistry	Ligand functional group	Commercial example
Cyanogen bromide	$-NH_2$	CNBr-activated Sepharose (Pharmacia)
Tresyl chloride	$-NH_2$, $-SH$	TSKgel Tresyl-5PW (TosoHaas Corporation)
Glycidoxypropyltrimethoxy silane	$-NH_2$	PROSEP-5CHO (Bioprocessing)
Aminopropyltriethoxy silane	$-NH_2$	PROSEP-9CHO (Bioprocessing)
Divinyl sulphone	$-NH_2$	Mini-Leak (Kem-En-Tec)
N-hydroxysuccinimide	$-NH_2$	Affigel 10 (Bio-Rad)

2.5 Small ligands and spacer options

When a small ligand such as a peptide is coupled to a matrix, a major steric problem can be encountered. The problem relates to the local steric interference generated by the matrix on the microenvironment of the immobilized ligand. A small ligand attached directly to the solid support may not protrude sufficiently far to interact with the binding site of an approaching protein (e.g. an antibody) in the sample. The problem is solved by placing the ligand at the terminus of a long chain or 'spacer arm' (2, 6).

The 'spacer arm' should be a chain of carbon atoms or indeed a protein. The choice of the spacer molecule can affect the relative hydropathicity of the immediate environment of an immobilized ligand. For example, molecules containing long hydrocarbon chains may increase the potential for non-specific binding by hydrophobic interaction. It is also possible to use even a large protein such as bovine serum albumin as a spacer arm to which several molecules of a small ligand can then be coupled. Alternatively, the avidin–biotin system can be exploited: peptide is first conjugated to biotin which is then bound to avidin which is covalently immobilized on the matrix. The following molecules have been employed as spacer groups to distance a ligand from a matrix (numbers in brackets indicate the number of carbon atoms in the spacer arm):

- ethylene diamine [2]
- 1,3-diamino-2-propanol [3]
- succinic acid [4]
- 6-aminocaproic acid [6]
- 1,6 diaminohexane [6]

2.6 Protein A bead-linked antibodies

The binding of monoclonal antibodies to matrix-linked Protein A is usually exploited for immunoglobulin and antibody purification (Section 5). However, it can be employed to produce an immunoadsorbent containing antibody covalently linked to the matrix-bound Protein A (7). In this case, the Protein A may be considered to be acting as a macromolecular spacer with the added advantage of orientating antibody binding sites in a controlled manner towards the mobile fluid phase. However, unoccupied Protein A molecules remaining after coupling will be available to bind contaminating immunoglobulins in the sample.

Essentially, Sepharose–Protein A beads, loaded with the antibody of interest, are suspended in 10 volumes of 0.2 M sodium borate, pH 9.0, and solid dimethylpimelimidate is added to bring the concentration to 20 mM. After mixing for 30 min at room temperature, the reaction is stopped by washing and the beads are incubated with 0.2 M ethanolamine, pH 8.0, for 2 h. The beads can then finally be washed into phosphate-buffered saline (PBS) before use.

3. Chromatographic procedures

The principle of immunoaffinity chromatography is a simple one involving:

- sample application to the column
- adsorption of product to the matrix
- removal of contaminants
- elution of product

It is unnecessary to devise a complex experimental protocol to achieve successful separations although there are a number of points that should be considered.

3.1 Equipment

The basic hardware for immunoaffinity chromatography should include the chromatographic column containing the matrix, a pump to apply the sample and washing and elution buffers, a UV or other appropriate detector system to monitor the eluant from the column, and a fraction collector to collect the product. The dimensions of the separation column are not critical to the efficiency of the process but a short squat column permits greater versatility in the selection of flow rates that may be achieved during the experiment.

The basic apparatus required is illustrated in *Figure 1* which also incorporates several extra features designed to enhance the ease of operation of the procedure. A guard column containing underivatized matrix material can be placed in line before the separation column. This offers protection to the separation column containing the immunoadsorbent matrix which, under optimal circumstances, can be used many times with little or no loss of activity. The guard column provides an additional line of defence for the elimination of extraneous materials (e.g. protein aggregates, air-borne dust particles) which may contaminate the sample prior to its fractionation.

A third column containing a gel filtration medium such as Sephadex G25 (Pharmacia) can be placed in-line following the separation column. This column is by-passed during sample application and only connected in series prior to applying the dissociating agent used to desorb the product from the matrix. Thus, as the agent passes through the separation column, releasing product, it reaches the gel filtration column which has been selected to fractionate the product (a macromolecule of higher molecular weight, which elutes in the void volume) from the dissociating agent (of low molecular weight, which is retarded by the gel filtration medium). If no gel filtration column is included in the apparatus, the eluate from the separation column can be collected directly into tubes containing an agent to neutralize the potentially denaturing desorbing agent, for example 1 M Tris base (pH 8), to

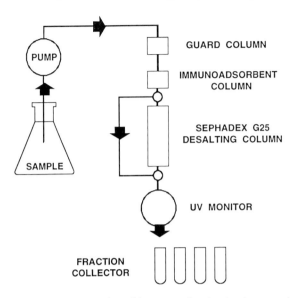

Figure 1. Diagrammatic representation of immunoadsorbent column and ancillary apparatus used for immunoaffinity chromatography.

neutralize acid eluates. However, complete mixing of eluate with neutralizing agent will not occur without additional shaking of each collecting tube.

The fractionation procedure can employ a variety of in-line, monitoring processes (UV, conductivity, pH, etc.) to check on the efficacy of the separation, and the whole process can be easily automated with relatively simple additional control units. However, even though fractionation procedures may be subjected to sophisticated control, this does not guarantee a better product than that which can be obtained using a simple 'open-top' immunoadsorbent column, a manual pipette, and careful experimental technique. The key to successful immunoaffinity chromatography relates primarily to the very nature of the antigen–antibody interaction being exploited to effect the separation.

Finally, some thought should be given as to whether it is necessary to set the apparatus up in a cold-room or in the general laboratory. The determining factor here is stability of sample, the column matrix material, and isolated product, although if there is any doubt, caution would direct the work towards the cold-room.

3.2 Sample preparation and application

Several questions should be considered before applying a sample to an immunoaffinity column:

(a) Is the column ready to receive a sample? It is good practice to flush the column through with the dissociating agent to be used later in the separation and then equilibrate with the washing buffer.

(b) Has all particulate matter (aggregates) been removed from the sample? The sample to be applied to the column must be subjected to ultrafiltration (e.g. using a 0.2 μm filter) or ultracentrifugation (typically 40 000g for 60 min) to ensure that all particulate matter has been eliminated.

(c) Is it necessary to add a preservative to the sample? With large sample volumes and small columns the sample may be standing for many hours or even days while waiting to be passed through the affinity column. Sodium azide (0.05%) or enzyme inhibitors such as phenylmethylsulphonyl chloride (50 μg/ml) may be included to ensure that the sample does not deteriorate (see also Section 6.2) or alternatively the use of a Tris-based buffer system (10 to 20 mM) is bacteriostatic over a short period.

(d) Does the flow rate of the pump allow sufficient time for the product in the sample to interact with the matrix? This will be determined by experiment, but as a starting point all of the material in the sample should be allowed at least 30 min to interact with the column matrix. One option here is to recirculate the sample solution through the column over a convenient period (e.g. overnight).

The volume of the sample to be applied is not a major problem. Immunoaffinity chromatography has the capacity to concentrate the product from a very dilute solution by specific adsorption to the substituted matrix, and indeed this is an important characteristic of the procedure.

3.3 Column washing

After application of the sample to the column, the column is washed initially with the sample diluent solution such as PBS or 20 mM Tris-buffered saline (TBS) at neutral pH in order to eliminate material which is not bound to the affinity matrix. Washing of the column matrix is continued until the UV monitor shows a stable baseline (usually after five to ten column bed volumes). However, some contaminants in a complex, highly heterogeneous sample may interact with the matrix and adsorb non-specifically (by processes other than the intended antigen–antibody interaction). These interactions would include hydrophobic binding to matrix-bound material, ionic interactions, and conformational occlusion phenomena. To overcome these interfering effects various pre-elution treatments have been attempted including a pH shift of the wash buffer or high ionic strength treatment before elution of specifically bound product from the column. If a pre-elution step is considered necessary to release non-specifically bound materials, then the following buffer systems may be tried:

- 1 M to 2 M NaCl in 0.1 M phosphate buffer, pH 7.2
- 0.1 M acetate buffers of decreasing pH (± 1 M NaCl)
- 0.1 M glycine–NaOH buffers of increasing pH (± 1 M NaCl)

3.4 Column elution

A variety of agents are available to release an antigen or antibody from the washed immunoadsorbent matrix. These include:

- low pH (e.g. 0.1 M glycine–HCl, pH 2.8)
- high pH (e.g. 0.1 M diethylamine, pH 11.5)
- high salt (e.g. 3.5 M MgCl$_2$)
- chaotropic agents (e.g. 3 M NaSCN or KSCN)
- ionic detergents (e.g. 1% sodium dodecyl sulphate)
- denaturants (e.g. 2 M guanidine–HCl)
- organic solvents (e.g. 10% dioxane)

The product (antigen or antibody) should elute from the immunoadsorbent column as a single sharp peak. Once this has been recovered in the tubes of the fraction collector, the column should be returned to its starting buffer by pumping five to ten column volumes through the matrix. This buffer should contain a preservative (0.05% NaN$_3$ or 0.01% merthiolate) for long-term storage at 4 °C.

If the product (the released antigen or antibody) has not already been separated from the dissociating agent by passage through an in-line gel filtration column (*Figure 1*) or it has not been collected into tubes containing a neutralizing agent (Section 3.1), then tubes containing product should be identified and their contents pooled. The pool can then be dialysed against a neutral buffered salt solution (e.g. PBS) to separate the elution agent from the product.

It will be a matter of experiment to determine whether, and to what extent, the immunoadsorbent matrix survives exposure to the elution buffer system and time is well worth spending upon the selection of an agent which both elutes the bound product in an active form and leaves the column matrix in a reusable state. Clearly, if an antigen is not released from the affinity matrix using one of the milder treatments (pH change, high salt solution), then such solutions may be considered as pre-wash buffers to eliminate non-specifically bound material prior to antigen elution.

3.5 Quality control testing

As with any separative procedure, steps should be taken to monitor the recovery of the product (antigen or antibody) at every stage of its purification (8). This enhances the efficiency of the process by identifying procedures leading to loss of immune reactivity. Quantitative immunoassays for the product being purified permit calculation of recoveries and specific activities (activity units per microgram). Also, by titrating starting material and final product, a relative activity may be determined by comparing initial and final titres, and normalizing to the total amount in each fraction (8). Methods for the assay of peptides and their antibodies are given in Chapter 6.

In addition to estimating the immunological activity of the product isolated by immunoaffinity chromatography, it is also necessary to have an independent assessment of biochemical homogeneity, for example by sodium dodecyl sulphate polyacrylamide gel electrophoresis (SDS-PAGE) (9).

4. Antibodies

Antibodies, and particularly those of the IgG class, represent a major tool to the practising immunochemist. Their overall stability allows them to be enzymically split into well defined fragments, to be coupled to other molecules (enzymes, fluorescent dyes, drugs, toxins, etc.) and to radioisotopes, and this may be achieved with little or no loss in their capacity to bind antigen. Furthermore, they may be physically adsorbed to the surfaces of wells of microtitre plates to be used in immunoassays for antigen measurement, and they may be covalently linked to particulate matrices and packed into a column to be used for the purification of antigens by immunoaffinity chromatography.

4.1 Sources of antibodies

Serum from immunized laboratory animals will contain around 10 mg/ml of polyclonal IgG immunoglobulin, although even after hyperimmunization rarely will the specific antibody reach 10% of the total IgG fraction (i.e. a concentration of 1 mg/ml). In tissue culture supernatants of hybridomas producing monoclonal antibodies, specific antibody may be present at concentrations up to 50 µg/ml, although care should be taken to culture the hybridoma in a medium which contains little or no contaminating immunoglobulin from the serum supplement. In this case, serum-free media or medium supplemented with fetal bovine serum should be used. Ascitic fluids produced in response to hybridoma cells growing in the peritoneal cavity of mice may contain high concentrations of specific antibody (up to 5 mg/ml) although normal mouse immunoglobulins will be a major contaminant and difficulties will be encountered in obtaining pure monoclonal antibody.

4.2 Storage and stability

Antibodies in serum or ascitic fluids are relatively stable to long-term storage at −20 °C or below. However, since repeated freezing and thawing causes aggregation and a loss in activity, storage in appropriate aliquots is recommended. Aggregates should be removed before use by centrifugation (10 000 g for 5 min using a bench microcentrifuge is usually sufficient) for immunoassay work, but samples for fractionation should satisfy more rigorous criteria for solubility (Section 3.2).

Monoclonal antibodies in tissue culture supernatants are usually stable to storage at 5 °C for periods of several months and longer provided that microbial contamination is minimized by the addition of NaN_3 to 0.05%, and the pH is adjusted to neutrality, if necessary.

Purified antibodies should be stored in isotonic solution at neutral pH (e.g. in PBS) in aliquots at −20 °C or with 0.05% NaN_3 at 5 °C. However, since NaN_3 interferes with many coupling reactions, it may be necessary to remove it by dialysis before further use. Antibody concentrations of 1 mg/ml or greater are recommended.

5. Antibody purification

Techniques for antibody purification include the broad range of conventional methods involving precipitation and column chromatography (10), although none generally produce the degree of purity achieved with relative ease using affinity techniques.

Antibody concentrations are determined by UV absorbance at 280 nm assuming that 1 mg/ml solutions of IgG and IgM give measurements of 1.43 and 1.19 respectively. (These values for mouse immunoglobulin are appropriate for Igs of other species.) Antibody purity is checked by analysis of a sample by SDS-PAGE. A sample of the antibody solution is prepared under reducing and non-reducing conditions so that intact immunoglobulin and immunoglobulin subunits (heavy and light chains respectively), can be identified. Sample overloading permits low-level contaminants to be detected.

5.1 IgG monoclonal antibodies

Murine IgG monoclonal antibodies in tissue culture supernatants and ascitic fluids may be most conveniently purified by their binding to, and elution from, Sepharose-linked Protein A or G (Pharmacia) or Protein A or G linked to CPG granules, for example PROSEP A (Bioprocessing Ltd), which provide matrices with a high capacity for binding mouse immunoglobulins. Simple affinity chromatography with acid (0.1 M glycine–HCl, pH 2.8) elution of antibody from the column, frequently yields antibody preparations of greater than 95% purity.

Particulate material (hybridoma cell debris) is first eliminated from the sample by centrifugation (typically, in our laboratory, $40\,000\,g$ for 60 min) and ascitic fluids are diluted 1/10 with PBS before application to the column.

5.2 IgG from serum

Proteins A and G have different profiles of affinities for IgG in serum of different animal species. *Table 3* gives a guide to the relative binding affinity of IgGs of a range of animal species for immobilized Protein A and Protein G. In addition, Proteins A and G have varying affinity for the different IgG

Table 3. Relative affinity of different animal IgG immunoglobulins and some IgG subclasses for Protein A and Protein G

Immunoglobulin	Affinity for[a]	
	Protein A	Protein G
Mouse IgG	+	+
Rat IgG	±	+
Rabbit IgG	+	+
Guinea pig IgG	+	±
Goat IgG	±	+
Sheep IgG	±	+
Chicken IgG	−	−
Mouse IgG1	±	+
IgG2a	+	+
IgG2b	+	+
IgG3	+	+
Human IgG1	+	+
IgG2	+	+
IgG3	−	+
IgG4	+	+

[a] '+' = strong binding; '±' = weak binding; '−' = non-binding.

subclasses in whole serum and the reactivity of different subclasses for human and mouse serum is also summarized in *Table 3*.

The varying affinity for the different mouse IgG subclasses for Protein A may be exploited to separate the individual subclasses. Protein A has the lowest affinity for IgG1 although adjusting the sample pH to 8.0 promotes IgG1 binding. After washing non-bound material from the column, bound antibody may be eluted with 0.1 M glycine–HCl, pH 2.8, or 3 M NaSCN. When fractionating mouse serum or ascitic fluid, individual subclasses of mouse immunoglobulin may be selectively desorbed from the Protein A matrix by sequential elution with 0.1 M citrate buffers of increasing acidity (11) as follows:

- pH 6.0 for IgG1
- pH 4.5 for IgG2a (and the minor subclass, IgG3)
- pH 3.5 for IgG2b

Clearly, the volume of each wash solution will depend upon the scale of the fractionation, but as a general rule each wash cycle with buffer should be at least five times the volume of the column.

Protein A linked to different matrices can be used for repeated fractionations. However, it is an expensive reagent, and in this case a guard column containing unsubstituted matrix material placed in series immediately before the affinity column is very worthwhile (*Figure 1*).

5.3 IgM monoclonal antibodies

For the purification of murine IgM antibodies, Sepharose-linked lentil lectin (12) and Sepharose-linked protamine sulphate (11) have been employed as affinity matrices for IgM isolation.

5.4 Antigen affinity chromatography

The major advantage of antigen affinity chromatography is that it facilitates the purification of specific antibody from a polyclonal antiserum. With a sample containing monoclonal antibody, purification on an antigen affinity column would be generally unnecessary, unless the sample was contaminated with irrelevant and/or unreactive immunoglobulin protein. Antigen is coupled to an appropriate insoluble chromatographic matrix using methods detailed in Section 2 and standard immunoaffinity chromatography is performed.

The major disadvantage with the system is that relatively large amounts of antigen may be required to be committed to the preparation of the matrix and there may be significant antigen loss or inactivation during usage (7).

5.5 Epitope affinity chromatography

Epitope affinity chromatography is a specialized type of antigen affinity chromatography in which the epitope for a particular antibody to be purified is linked covalently to the chromatographic matrix (13). Increasingly, epitopes for monoclonal antibodies are being identified using a variety of methods, including novel strategies using peptide libraries (14–17; see Chapters 7 and 8), and in the case of linear peptide epitopes these may be then prepared synthetically. Alternatively, if the antibody (polyclonal or monoclonal) to be purified has been developed against a synthetic peptide, then this peptide can be exploited as the immobilized ligand for antibody purification. Furthermore, the orientation of the peptide to the matrix can be controlled by synthesis of peptide with an appropriately reactive terminus, for example a peptide containing a terminal cysteine can be linked simply to Activated Thiol-Sepharose 4B (Pharmacia) (13).

Protocol 1 gives a simple procedure for the purification of monoclonal antibodies using epitope affinity chromatography (13). The system has been particularly useful for the purification of anti-MUC-1 mucin antibodies which recognize determinants within the immunodominant hydrophilic region of eight amino acid residues in the mucin protein core (17). This sequence is part of the 20 amino acid sequence which is repeated multiple times in the mucin core (18). Another procedure for the purification of anti-peptide antibodies is given in Chapter 6, *Protocol 9.*

Protocol 1. Purification of antibodies by epitope affinity
chromatography [a]

Equipment and reagents

- Fraction collector, pump, UV-monitor (see Figure 1)
- Two chromatography columns, 1.6 cm diameter
- CNBr-activated Sepharose (Pharmacia)
- Sephadex G25 (Pharmacia)

- Coupling buffer: (0.1 M NaHCO$_3$, pH 8.0, 0.5 M NaCl)
- PBS (20 mM sodium phosphate, pH 7.3, 0.15 M NaCl)
- 3 M NaSCN

Method

1. Substitute CNBr-activated Sepharose 4B with peptide according to the manufacturer's instructions. Typically, add peptide (e.g. A P D T R P A P G (13)) dissolved at 1 mg/ml in coupling buffer, at 2 mg of peptide per ml of activated gel.

2. Prepare an immunoadsorbent column (1.6 cm diameter) containing 5 ml of packed Sepharose-peptide and wash it with 25 ml of 3 M NaSCN followed by PBS.

3. Apply the hybridoma tissue culture supernatant, ascitic fluid (diluted 1/10 in PBS), or serum from immunized animals (diluted 1/10 in PBS) (prepared according to Section 3.2) to the column at approximately 20 ml/h. Fractions can be collected to check if, or at what stage, the column is fully saturated.

4. Upon completion of sample application wash the column with PBS (50 ml).

5. Connect the column in series to a Sephadex G25 column (1.6 × 15 cm equilibrated with PBS) and elute the antibody bound to the immuno-asdorbent column by application of 3 M NaSCN. (As the antibody elutes in the NaSCN front, and reaches the Sephadex column, it is immediately desalted into PBS.)

6. Pool the fractions containing antibody, dialyse against PBS to remove any residual NaSCN and clarify by ultracentrifugation or ultrafiltration to eliminate aggregates before storage (Section 4.2).

[a] This method links the peptide to the matrix via amino groups. When peptides are available with a cysteine as the N-terminal residue, the peptide may be immobilized to Activated Thiol-Sepharose 4B (Pharmacia). In this case, the peptide is dissolved at 1 mg/ml in 0.1 M Tris–HCl, pH 7.5, and conjugation to the matrix is performed according to the manufacturer's instructions using 2 mg peptide per ml of gel. Another method of coupling cysteine-containing peptides to agarose is described in *Protocol 9* of Chapter 6.

The capacity of epitope affinity matrices for binding antibody has proved to be extremely high. With the immobilized A P D T R P A P G peptide, up to

40 mg of an IgG antibody was bound per ml of gel and the capacity was around 10 mg/ml gel for an IgM antibody (which presumably because of its larger size, was unable to penetrate the matrix as efficiently as the IgG antibody) (13). It is interesting to note that with these high-capacity separations, the immunoadsorbent column actually changes colour as it becomes saturated—the saturated gel gains a more intense white coloration. Also, antibody concentrations in the eluate may be very high (reaching 10 to 20 mg protein per ml) so that problems of aggregation may be encountered.

5.6 Affinity purification with biotinylated peptides

An alternative procedure for immobilizing peptides on an affinity matrix for antibody purification has recently been described (19) and involves the use of biotinylated acrylic beads (Sigma). Briefly, the beads are saturated with avidin and this matrix is used to bind biotinylated peptide. These beads can then be used to capture specific anti-peptide antibodies which can be eluted with a high- or low-pH buffer (under conditions which do not dissociate the biotinylated peptide from bound avidin or avidin from the beads).

6. Antigen purification

The procedure adopted for antigen purification using an anti-peptide antibody immobilized on an affinity matrix will depend upon the nature and solubility of the antigen. If the antigen in question is a membrane-associated product then measures must be taken to render it soluble before proceeding with any separation. On the other hand, if the antigen is already a soluble product, this presents a simpler fractionation problem.

Membrane-associated proteins and glycoproteins may be categorized as:

- integral (intrinsic) membrane proteins which span or are anchored within the lipid bilayer
- peripheral (extrinsic) proteins which are more loosely associated with the surface, interacting with the bilayer by means of weak electrostatic forces

Integral membrane proteins are bound by hydrophobic interactions, and require detergents for extraction, whereas peripheral membrane proteins may be released by treatments which leave the integrity of the membrane intact. The prime consideration in the selection of a procedure for solubilization of antigen is that antigenic activity is preserved. Given this precondition, the complete range of procedures for extracting an antigenic membrane protein from its cellular environment are available (20).

6.1 Preliminary considerations

Before deciding to proceed with antigen purification it is worth considering the logistics of the exercise. In particular, the scale of the experiment needs to be defined. If, for example, a membrane protein of, say, 60 kDa is to be isolated and there are 10^5 copies per cell, then to obtain a few micrograms of purified antigen (e.g. 10 μg), a total of at least 10^9 cells would be required as well as 100% recovery of product at each stage of the fractionation! While it is feasible to generate such cell numbers from tissue culture, efficiencies at this level cannot be achieved. Alternatively, with many other antigens, products may be present in greater concentrations or the material for extraction is available in considerably greater quantities. For example, we purify carcino-embryonic antigen from acid extracts of 1 kg quantities of liver metastases and, even using conventional chromatographic fractionations, around 100 to 200 mg of antigen is isolated per kilogram of tissue. However, whatever the problem, in the ideal case the recovery of antigenic activity should be accurately assessed at each stage in the separation, in order to maximize the efficiency of the fractionation procedure.

6.2 Storage and stability

The general comments made concerning storage and stability of antibodies (Section 4.2) are relevant to a variety of protein and glycoprotein antigens. Again, it is highly desirable to develop a quantitative assay for the antigen to be purified since this will permit appropriate conditions for storage to be defined. A major problem causing the loss of antigenic activity during storage is due to degradation by proteases in the often complex sample to be fraction-ated. If this is so, an appropriate enzyme inhibitor should be added to alleviate the specific problem or a cocktail of inhibitors can be added as a preventative measure. The following are a selection of the more commonly used protease inhibitors:

- aprotinin (serine protease inhibitor)—add to a final concentration of 0.1 μg/ml from a 10 mg/ml stock solution in PBS
- leupeptin (serine and thiol protease inhibitor)—add to a final concentration of 1.0 μg/ml from a 10 mg/ml stock solution in H_2O
- pepstatin (acid protease inhibitor)—add to a final concentration of 1.0 μg/ml from a 10 mg/ml stock solution in methanol
- phenylmethylsulphonyl fluoride (serine protease inhibitor)—add to a final concentration of 50 μg/ml from a 10 mg/ml stock solution in isopropanol

6.3 Purification of soluble antigens

Purification of soluble antigens by immunoaffinity chromatography presents no problems in general other than those imposed by the reagents, and par-ticularly the antibody or polyclonal antiserum used to prepare the immuno-adsorbent column. The strength of the union between the antigen and the immune matrix is perhaps the major factor requiring special attention. Pre-liminary tests are worth conducting to determine the optimal conditions for antigen elution and as outlined in Section 3.4, the aim is to identify the mildest form of elution which will release antigen from the antibody with ease and with no loss of antigenic activity, and also leave the immunoadsorbent column in a reusable condition. Acidic and basic buffers should be tried first, then high salt, chaotropes, denaturants, and finally ionic detergents (Section 3.4).

6.4 Purification of membrane-associated antigens

The purification of cell membrane-associated antigens may be carried out using affinity chromatography with immobilized antibodies, but first the antigen must be rendered soluble in the buffers to be used for the fractionation. This may be achieved by a variety of means including: limited proteolysis to release an antigenic fragment, extraction with acid treatment, high salt, chaotropic agents, or organic solvents (e.g. *n*-butanol), physical treatments such as ultrasonication, and treatment with detergents. The last is perhaps the most widely employed procedure to effect antigen release prior to purification and its particular merit is that a native form of the antigen is efficiently solubilized. A variety of detergents are available for antigen solubilization (12) and again the final choice will depend much upon trial and error, and the compromise will be between increasing the efficiency of solubilization and retaining antigenic activity. In addition, having selected the detergent for solubilization, care has to be taken to ensure that the chromatography buffers will maintain antigen solubility throughout the subsequent separation. In practice, this simply means that low concentrations of detergent should be included in the chromatography buffers used in the separation.

The following protocol has found wide application for the purification of proteins from tumour extracts solubilized by treating subcellular membranes with the non-ionic detergent Nonidet NP-40 (0.2%) (21–23). These extracts are highly heterogeneous but even so the efficiency of the immunoadsorbent chromatographic procedure facilitates the achievement of considerable purification for relatively minor membrane-associated antigens.

Protocol 2. Purification of membrane-associated proteins by antibody affinity chromatography

Equipment and reagents

- Fraction collector, pump, UV monitor, chromatography columns (see *Figure 1*)
- Tissue homogenizer (e.g. rotating blade type or rotating close-fitting pestle type)
- Centrifuge and ultracentrifuge
- CNBr-activated Sepharose (Pharmacia)
- 0.1 M diethylamine, pH 11.5
- 0.5% NP-40 in 0.1 M Tris–HCl, pH 7.0
- 0.5% NP-40 in 0.1 M Tris–HCl, pH 7.6
- 0.2% NP-40 in 0.1 M Tris–HCl, pH 7.6, 1 M NaCl
- 0.1% NP-40 in 0.1 M Tris–HCl, pH 7.6
- 1 M Tris–HCl, pH 7.6
- PBS

Protocol 20. *Continued*

Method

1. Substitute CNBr-activated Sepharose with antibody according to the manufacturer's instructions. Routinely, a coupling efficiency of greater than 95% should be expected. Substitute the gel with antibody protein added in the range 0.5 to 5 mg per ml of gel.

2. Pack the immunoadsorbent into a chromatography column and pre-wash the column with the agent (e.g. 0.1 M diethylamine, pH 11.5) to be used for antigen elution and equilibrate with 0.1% NP-40 in 0.1 M Tris–HCl, pH 7.6.

3. Prepare an homogenate and remove nuclei and whole cells by centrifugation (600 *g* for 10 min); sediment a total subcellular membrane preparation by centrifugation (100 000 *g* for 60 min) and discard the cytosol supernatant.

4. Disperse the membrane pellets in 0.5% NP-40 in 0.1 M Tris–HCl, pH 7.0 (added at 5 ml per gram of original tissue) and extract with stirring for 30 min at 5 °C.

5. Clarify the extract by centrifugation at 100 000 *g* for 60 min and pump on to the column. For convenience, the extract may be recycled continuously through the column.

6. Wash the column[a] successively with:

   • 0.5% NP-40 in 0.1 M Tris–HCl, pH 7.6

   • 0.2% NP-40 in 0.1 M Tris–HCl, pH 7.6, 1 M NaCl

   • 0.1% NP-40 in 0.1 M Tris–HCl, pH 7.6

7. Apply 0.1 M diethylamine, pH 11.5, to the column to elute the antigen. Fractions are collected into tubes containing sufficient 1 M Tris–HCl, pH 7.6, to adjust the pH of the fraction to less than 8.0.

8. Identify antigen-containing fractions using an appropriate immunoassay. Pool those fractions containing antigen, dialyse against PBS (and concentrate if necessary) and store in small aliquots at − 20 °C. If any aggregation has occurred during this latter processing, then insoluble material should be removed by ultracentrifugation before storage.

9. Analyse the antigen preparation for homogeneity, e.g. by SDS–PAGE (9), and antigenic activity.

---

[a] The volume of each wash solution will depend upon the scale of the fractionation, but as a general rule each wash cycle with buffer should be at least five times the volume of the column.

Protocol 2 was originally developed for the purification of gastrointestinal tumour membrane-associated antigens (21) and subsequently the method has been applied to the purification of breast and ovarian tumour membrane-associated mucins (22, 23). In these cases, it was subsequently determined using epitope mapping tests that the antibody used to prepare the immuno-adsorbent recognized the trimeric sequence, RPA, in the mucin protein core (17). Similar mucins have been isolated from urine (these mucins are exfoliated from the urothelium) (24) or from breast cancer patients' serum (having been shed into the circulation from the tumour) (25). However, since urine and serum are sources of water-soluble antigen, it was unnecessary to include the non-ionic detergent, NP-40, in the various chromatography buffers in *Protocol 2*. Recently, monoclonal antibodies have been developed against a synthetic peptide with a sequence based upon that of a gastrointestinal mucin and these anti-peptide antibodies have been employed using *Protocol 2* for the isolation of mucin from human colonic tumour (26).

7. Conclusions

With the development of monoclonal antibodies, immunoaffinity chromatography has become a major technique for the purification of antigenic molecules from complex, multicomponent systems. The technique not only has the capacity to effect considerable purification, but also the antigenic material eluted from the column is concentrated during the separation procedure. The methodology has been widely exploited in the molecular sciences, and since it is a relatively simple process to scale up the technique, immunoaffinity chromatography is used increasingly in many biotechnological areas and particularly for the separation of products of fermentation.

The identification of the epitopes of many antibodies is now feasible using various mapping procedures (14–17; see Chapters 7 and 8). In addition, anti-peptide antibodies (which are also reactive with the parent protein) are being produced for probing biological systems (see Chapter 4). Consequently, epitope affinity chromatography, using immobilized peptide epitopes, is becoming applicable in an increasing number of systems for antibody purification. This approach contributes to the retention of maximum immunoreactivity of the antibody product—denatured antibodies which are no longer functionally active are simply washed out with the contaminants. This is particularly important if the antibody is to be used for diagnostic imaging or therapeutic manipulations in patients.

Finally, as a technical note, attention has been focused upon the use of immunoadsorbent matrices for chromatographic fractionations, rather than batch-wise separations. Clearly, affinity matrices can be used in both approaches. However, for most purposes, column fractionations allow greater control of sample application and product elution. Elution of product in a batch process will require harsher eluting buffers (to reduce rebinding) and promote greater dilution of the product than can be achieved on a column.

Acknowledgements

These studies were supported in part by the Cancer Research Campaign (CRC). MRP is a CRC Reader in Pharmaceutical Sciences and his research is supported by Project Grant Number SP 2168/0101 from the Cancer Research Campaign.

References

1. Kohler, G. and Milstein, C. (1975). *Nature*, **256**, 495.
2. Lowe, C. R. (1979). *An introduction to affinity chromatography*. North-Holland, Amsterdam.
3. Gleick, J. (1988). *Chaos*. Sphere Books, London.
4. Groman, E. V. and Wilchek, M. (1987). *Trends Biotechnol.*, **5**, 220.
5. Taylor, R. F. (1985). *Anal. Chim. Acta*, 241–8.
6. Dean, P. D. G., Johnson, W. S., and Middle, F. A. (1985). *Affinity chromatography: a practical approach*. IRL Press, Oxford.
7. Harlow, E. and Lane, D. (1988). *Antibodies: a laboratory manual*. Cold Spring Harbor Press, Cold Spring Harbor, NY.
8. Williams, A. F. and Barclay, A. N. (1986). In *Handbook of experimental immunology. Part 1. Immunochemistry,* 4th edn, (ed. D. M. Weir), pp. 22.1–22.17. Blackwell Scientific Publications, Oxford.
9. Laemmli, U. K. (1970). *Nature*, **227**, 680.
10. Stanworth, D. and Turner, M. W. (1986). In *Handbook of experimental immunology. Part 1. Immunochemistry*, 4th edn, (ed. D. M. Weir), pp. 12.1–16. Blackwell Scientific Publications, Oxford.
11. Hudson, L. and Hay, F. C. (1989). *Practical immunology*, 3rd edn. Blackwell Scientific Publications, Oxford.
12. Takacs, B. and Staehelin, T. (1981). In *Immunological methods* (ed. I. Lefkovits and B. Pernis), Vol. 2, pp. 27–56. Academic Press, New York.
13. Price, M. R., Sekowski, M., and Tendler, S. J. B. (1991). *J. Immunol. Methods*, **139**, 83.
14. Geysen, M., Rodda, S. J., Mason, T. J., Tribbick, G., and Schoofs, P. G. (1987). *J. Immunol. Methods*, **102**, 259.
15. Lam, K. S., Salmon, S. E., Hersh, E. M., Hruby, V. J., and Kazmierski, W. M. (1991). *Nature*, **354**, 82.
16. Houghten, R. A., Pinilla, C., Blondelle, S., Appel, J. R., Dooley, C., and Cuervo, J. H. (1991). *Nature*, **354**, 84.
17. Price, M. R., Hudecz, F., O'Sullivan, C., Baldwin, R. W., Edwards, P. M., and Tendler, S. J. B. (1990). *Mol. Immunol.*, **27**, 795.
18. Gendler, S., Taylor-Papadimitriou, J., Duhig, T., Rothbard, J., and Burchell, J. (1988). *J. Biol. Chem.*, **263**, 12820.

19. Affinity purification of antibodies with biotinylated peptides. *Pinnacles* (1992) **2**(3), 3–4.

20. Price, M. R. (1993). In *Immunobiology: a practical approach* (ed. G. Gallagher, R. C. Rees, and C. W. Reynolds), pp. 59–79. IRL Press, Oxford

21. Blaszczyk, M., Pak, K. Y., Herlyn, M., Lindgren, J., Pesano, S., and Koprowski, H. (1984). *Cancer Res.*, **44**, 245.

22. Price, M. R., Edwards, S., Powell, M., and Baldwin, R. W. (1986). *Br. J. Cancer*, **54**, 393.

23. Price, M. R., Edwards, S., Owainati, A., Bullock, J. E., Ferry, B., Robins, R. A., and Baldwin, R. W. (1985). *Int. J. Cancer*, **36**, 567.

24. Price, M. R., Pugh, J. A., Hudecz, F., Griffiths, W., Jacobs, E., Symonds, I. M., Clarke, A. C., Chan, W., and Baldwin, R. W. (1990). *Br. J. Cancer*, **61**, 681.

25. O'Sullivan, C., Price, M. R., and Baldwin, R. W. (1990). *Br. J. Cancer*, **61**, 801.

26. Durrant, L., Jacobs, E. and Price, M. R. (1994). *Eur. J. Cancer.*, **30A**, 355.

Immunoassays of proteins and anti-peptide antibodies

NIGEL P. GROOME

1. Introduction

The use of antibodies in the qualitative and quantitative detection of proteins and peptides is well established. In this chapter special emphasis will be placed on the detection of antibodies reacting with peptides, and the application of monoclonal and polyclonal antibodies raised using synthetic peptide immunogens to the detection of the parent protein. Examples will be given of cases where anti-peptide antibodies have been successfully applied, and an attempt made to give the reader a balanced view of some of the advantages and disadvantages of this approach. Several points should be made clear at the outset:

(a) Most immunoassay, immunocytochemistry, and immunoblotting techniques are carried out in an identical fashion whether the antibodies have been raised to peptides or not.

(b) In this chapter no attempt at comprehensive coverage will be made. Instead, specific techniques which have been used in the author's laboratory will be described in detail, and an attempt made to share some of the experiences in the applications of anti-peptide antibodies.

1.1 Why use anti-peptide antibodies?

Several examples can be given of circumstances where anti-peptide antibodies have advantages over those raised to the parent protein:

(a) If the amino acid sequence of a low-abundance protein of interest has been deduced from a nucleic acid sequence the application of synthetic peptides may be the only immediate route to the preparation of useful antibodies. This is likely to become of increasing importance as more nucleic acid sequences become available for proteins of unknown function.

(b) Antibodies made using synthetic peptides have the advantage that they are of pre-determined specificity, i.e. it can be assumed that they react with the corresponding protein at a defined point along the primary sequence. Such antibodies may be useful for topographical analysis of functional sites in hormones, enzymes, and receptors. In an ideal situation an antibody to one region of the molecule unambiguously blocks one function while leaving another unaltered.

The advantage of predetermining the epitope to be recognized is critical when the need is for antibodies which detect specifically each of several highly similar molecules. Recently anti-peptide antisera were developed which were capable of distinguishing three isoforms of transforming growth factor β by immunizing with peptide sequences from one of the few variable regions of the molecule (1). A similar approach for distinguishing multiple isozymes of protein kinase C has been made using antibodies to peptides from the variable C-terminal region (2). Tsitsiloni *et al.* (3) established an anti-peptide radioimmunoassay (RIA) for rat parathymosin, which did not cross-react with closely related thymosin and prothymosin, by using peptides from the variable N-terminal as tracer and immunogen. Duncan *et al.* (4) used anti-peptide monoclonal antibodies to develop a specific enzyme-linked immunosorbent assay (ELISA) for the neuron-specific γ isozyme of human enolase by targeting a sequence where the γ and α isozymes differed.

(c) High-affinity antibodies to closely conserved molecules may be difficult to make by immunization with natural or recombinant material because of strong self-tolerance in the host immune system. Examples may be given of situations where an appropriate peptide has been coupled to a carrier to produce useful antibodies where previous attempts with purified protein had failed. In the author's own work, this applied to the preparation of antibodies to the conserved βA subunit of inhibins and activins, where a peptide sequence from near the C-terminal induced the first high-affinity monoclonal for this molecule suitable for use in immunoassay (5, 6).

(d) Using synthetic peptides, antibodies can be made to neoepitopes created at the ends of peptide fragments generated by proteolysis, and which are not present on the parent molecule. Examples are the use of an N-terminal sequence from a fibrinopeptide to generate antibodies specific for fibrin (7), and the author's use of anti-peptide antibodies to detect small fragments arising from the proteolytic cleavage of myelin basic protein (8).

(e) Antibodies raised to synthetic peptides are more likely to recognize the target molecule after denaturing treatments than many antibodies made by conventional means, which commonly recognize conformational epitopes. They are particularly valuable for immunoblotting after sodium dodecyl sulphate polyacrylamide gel electrophoresis (SDS-PAGE) and, depending on the amino acid sequence of the epitope, may be more resistant to the fixatives used for immunocytochemistry.

2. Immunoassays for the detection of antibodies

Three situations will be considered:

- The use of synthetic peptides attached to microplates for the detection of antibodies induced by natural or artificial immunization with native proteins.

- The use of synthetic peptides to detect antibodies resulting from immunization with the same or similar synthetic peptides.

- The use of the parent protein to detect antibodies raised with peptide immunogens.

2.1 Detection of antibodies to parent proteins which react with immobilized peptides

A good example of the use of solid-phase synthetic peptides in ELISA is the detection of antibodies to human immunodeficiency virus (9, 10). There is now a wide choice of methods available for attaching peptides to plastic, ranging from simple adsorption to covalent attachment in specific orientations. No method can be applied in every case and workers should be prepared to evaluate several methods for their particular application. Some of the methods of preparing solid-phase peptides for ELISA are as follows:

- simple adsorption to microplates (see *Protocol 1*)
- attachment to glutaraldehyde-treated microplates (see *Protocol 2*)
- attachment to adsorbed phenylalanine–lysine co-polymers (see *Protocol 3*)
- attachment to UV-treated microplates (see *Protocol 4*)
- attachment to Covalink microplates (see *Protocol 5*)
- adsorption of peptide–carrier protein conjugates (as for *Protocol 1*)
- adsorption of branched peptides (as for *Protocol 1*)
- attachment of N- or C-terminal biotinylated peptides to streptavidin-coated microplates (see *Protocol 8*)

Simple adsorption of peptides to standard ELISA plates from 50 mM carbonate/bicarbonate buffer (pH 9.6) overnight at 4°C is a commonly used procedure, and its simplicity commends it for initial attempts (*Protocol 1*). Very short peptides, however, may be lost excessively during normal ELISA wash stages leading to loss of precision. Also the conformation of the peptide may be so altered by its interaction with the plastic as to diminish the interaction with antibody. If one has preliminary ELISA data (based on simple adsorption) suggesting that a particular peptide can react with antibodies in a sample of interest it is quite likely that application of some of the other attachment methods may improve the sensitivity and precision of their detection.

141

Protocol 1. Simple adsorption of peptides and proteins to microplates[a,b,c,d]

Equipment and reagents

- ELISA-grade polystyrene microplates (e.g. Nunc 4–39454A or Dynatech M129A)
- PBS tablets (Sigma P4417). These give a solution of 10 mM sodium phosphate buffer, pH 7.4, 2.7 mM KCl, 0.137 M NaCl
- Bovine serum albumin (BSA), protease-free grade (Sigma A3294)
- PBS, 0.05% Tween 20 (PBST)

Method

1. Dissolve the peptide or protein in 50 mM carbonate buffer, pH 9.6, and dilute to a concentration of 10 µg/ml in the same buffer. (If the peptide is difficult to dissolve in this solution it may help to dissolve it in dimethylformamide and then dilute it to the required concentration. Acidic peptides often dissolve best in basic buffers and basic peptides best in acidic buffers. It is commonly observed that proteins and peptides adsorb most to plastic at a pH above their pI. Always give peptides plenty of time to dissolve and stand the sample in a sonicating bath if one is available.)

2. Add 100 µl of peptide solution to each well on a microplate. Cover the plate.

3. Incubate the plate overnight at 4 °C.

4. Shake dry and add 150 µl of PBS, 1% (w/v) BSA to block excess adsorption sites.

5. Incubate for 1 h at room temperature.

6. Wash the plate thoroughly with PBST and agitate gently in this solution for 1 h to remove loosely attached peptide.

7. Wash the plate again as in step **6** and use it immediately for ELISA (*Protocol 6*), or shake it dry, allow to thoroughly air dry, and store in a desiccated box at 4 °C.

[a] Some peptides may adsorb very weakly and require other methods of attachment.

[b] Some peptides may adopt a conformation unfavourable to antibody recognition.

[c] Peptides containing methionine, cysteine, or tryptophan may oxidize when attached to a surface, particularly if the plates are stored in air.

[d] There is an optimum coating concentration for each adsorbed peptide. If this is exceeded some peptide may be adsorbed very weakly, and if this desorbs during the assay, sensitivity and precision may be impaired. It is recommended that initial experiments be done to optimize the coating concentration.

Protocol 2. Attachment of peptides to glutaraldehyde-treated microplates[a,b]

Equipment and reagents

- Glutaraldehyde (25% w/v) (Sigma grade II G6257), preferably an older bottle which is more likely to contain polymers. It is these which adsorb best to the plastic
- ELISA grade microplates (Nunc 4-39454A or Dynatech M129B)

- BSA (*Protocol* 1)
- PBS (*Protocol* 1)
- PBST (*Protocol* 1)

Method

1. To each well of the plate add 100 μl of a 5% w/v solution of glutaraldehyde in PBS.

2. Incubate the plate for 1 h at room temperature.

3. Wash the plate thoroughly with clean PBS.[c]

4. Dilute the peptide in PBS to 2 μg/ml and add 100 μl to each well.

5. Incubate for 1 h at room temperature.

6. Shake dry and add 150 μl of PBS, 1% (w/v) BSA to block excess adsorption sites.

7. Incubate for 1 h at room temperature.

8. Wash the plate thoroughly with PBST and allow it to shake in this solution for 1 h to remove loosely attached peptide.

9. Wash the plate again as in step **8** and use it immediately for ELISA (*Protocol 6*), or shake it dry, allow to thoroughly air dry, and store in a desiccated box at 4 °C.

[a] This protocol is essentially that described by Kasprzyk *et al.* (11).
[b] At neutral pH the coupling is probably mainly through the N-terminal of the peptide. Coupling through the ϵ-amino side-chain of any lysines will increase with pH.
[c] It is important that fresh clean PBS is used so that no deactivation of the adsorbed glutaraldehyde takes place.

Protocol 3. Attachment of peptides to phenylalanine-lysine-treated microplates[a,b]

Equipment and reagents

- ELISA grade microplates (Nunc 4-39454A or Dynatech M129A)
- Phenylalanine-lysine 1:1 copolymer (Sigma P3150)
- *N*-acetylimidazole (Sigma A9625)

- Bis-(sulphosuccinimidyl) suberate (Pierce 21579)
- BSA (*Protocol* 1)
- PBS (*Protocol* 1)
- PBST (*Protocol* 1)

Protocol 3. *Continued*

Method

1. Dissolve 21 mg of co-polymer in 20 ml of distilled water. Stir the solution magnetically.

2. Add dropwise 1 ml of 0.2 M phosphate buffer, pH 7.5, to the stirred solution.

3. Dissolve 7.5 mg of *N*-acetylimidazole in 0.75 ml of dry acetone and add this dropwise to the co-polymer solution on the stirrer.

4. Stir for 5 h.

5. Dialyse the solution extensively against distilled water at 4 °C.

6. Centrifuge to remove any precipitate and filter the solution through a 0.22 μm membrane.

7. Check the concentration of the polymer from the absorbance at 257 nm. E 0.1% cm^{-1} = 0.535.

8. Store at −20 °C in aliquots.

9. Dilute the acetylated co-polymer stock solution in distilled water to 40 μg/ml and add 100 μl to each well of a microplate.

10. Cover the plate and incubate for 18 h at room temperature.

11. Wash the plate thoroughly with distilled water and shake it dry on a paper towel.

12. Add 100 μl of a freshly prepared 1 mg/ml solution of bis-(sulphosuccinimidyl) suberate in 10 mM phosphate buffer, pH 7.5, to each well.

13. Incubate at room temperature for 30 min.

14. Wash the plate thoroughly with 10 mM phosphate buffer, pH 7.5, and shake it dry.

15. Add 100 μl of peptide solution (10 μg/ml) in 10 mM phosphate buffer, pH 7.5.[c]

16. Incubate for 4 h at room temperature.

17. Wash the plate with 10 mM phosphate buffer, pH 7.5.

18. Block excess reactive sites with 10 mM ethanolamine–HCl, pH 8.6, for 30 min.

19. Shake the plate dry and add 150 μl of a PBS(') 1% (w/v) BSA to block excess adsorption sites.

20. Incubate for 1 h at room temperature.

21. Wash the plate thoroughly with PBS and allow it to shake in this solution for 1 h to remove loosely attached peptide.

22. Wash the plate again as in step **21** and use it immediately for ELISA (*Protocol 6*), or shake it dry, allow to thoroughly air dry, and store in a desiccated box at 4 °C.

[a] This protocol is essentially that described by Hobbs (12).
[b] At neutral pH the coupling is probably mainly through the N-terminal.
[c] Coupling with the ε-amino side-chain of lysine will increase with pH.

Protocol 4. Covalent coupling of peptides to UV-treated microplates[a,b]

Equipment and reagents

- ELISA grade microplates (Nunc 4–39454A)
- PBS (*Protocol 1*)
- PBST (*Protocol 1*)
- BSA (*Protocol 1*)

- UV irradiator TF-20C (Vilber Lourmat). This emits light of wavelength 254 nm and irradiates the plate at 3.39 J/cm^2
- BSA (*Protocol 1*)

Method

1. Irradiate the plate for 20 min with the open face of wells pointing towards the beam.

2. Immediately add 100 μl of peptide solution (25 μg/ml) in 0.1 M boric acid, 25 mM sodium tetraborate, 0.15 M NaCl, pH 8.4.

3. Cover the plate and incubate at 4 °C overnight.

4. Shake the plate dry and add 150 μl of a PBS, 1% (w/v) BSA to block excess adsorption sites.

5. Incubate for 1 h at room temperature.

6. Wash the plate thoroughly with PBST and allow it to shake in this solution for 1 h to remove loosely attached peptide.

7. Wash the plate again as in step **6** and use it immediately for ELISA (*Protocol 6*), or shake it dry, allow to thoroughly air dry, and store in a desiccated box at 4 °C.

[a] This protocol is essentially that described by Boudet *et al.* (13).
[b] Mechanism(s) of attachment are not known but may involve amino groups on the peptide (13).

Protocol 5. Attachment of peptides to Covalink microplates[a,b]

Equipment and reagents

- Nunc Immunomodules, Covalink NH (Nunc 478042)
- 1-ethyl-3-(3-dimethylaminopropyl) carbodiimide HCl (Sigma E6383)

- *N*-hydroxysuccinamide (Sigma H7377)
- BSA (*Protocol 1*)
- PBS (*Protocol 1*)
- PBST (*Protocol 1*)

Protocol 5. *Continued*

Method

1. Prepare a 4 mM solution of peptide in distilled water.

2. To 5 μl of peptide add 5 μl of aqueous 0.1M 1-ethyl-(3-dimethyl-aminopropyl)carbodiimide HCl and 5 μl of aqueous 0.1M N-hydroxy-succinamide.

3. Incubate for 30 min at room temperature.

4. Dilute the activated peptide to a concentration of 1 μg/ml in 0.1M carbonate/bicarbonate buffer, pH 8.6.

5. Add 100 μl of peptide solution to each well of a Nunc Covalink NH plate.

6. Incubate the plate for 2 h at 4°C.

7. Shake dry and add 150 μl of a PBS, 1% (w/v) BSA to block excess adsorption sites.

8. Incubate for 1 h at room temperature.

9. Wash the plate thoroughly with PBST and allow it to shake in this solution for 1 h to remove loosely attached peptide.

10. Wash the plate again as in step **9** and use it immediately for ELISA (*Protocol 6*), or shake it dry, allow to thoroughly air dry, and store in a desiccated box at 4°C.

[a] This protocol is essentially that described by Søndergard-Anderson *et al.* (14).
[b] Coupling is preferentially through the C-terminal and the side-chains of aspartic and gluta-mic acids.

2.2 Detection of antibodies to peptides which react with immobilized peptides

It should be borne in mind that a polyclonal population of antibodies raised to a synthetic peptide immunogen will contain antibodies recognizing the peptide in a wide range of different conformations. Each different solid-phase preparation of a peptide for ELISA has the potential to introduce bias into the detection. It would be possible for a particular peptide–carrier complex to induce apparently high levels of antibody measured by ELISA on plates prepared by one protocol, and apparently little antibody using another protocol. It should be emphasized that a good deal of caution is needed in making quantitative statements about anti-peptide antibodies measured by ELISA. With all solid-phase ELISAs only antibody above a certain threshold of affinity will be detected at each peptide coating density (15) and for polyclonal antibodies, ELISA is in no way a fully quantitative measure of total antibody.

2.3 Detection of antibodies to peptides which react with immobilized parent proteins

After the coating stage the ELISA procedures used to detect antibodies binding to peptides and purified proteins are identical. *Protocol 6* shows the procedure used in the author's laboratory.

Protocol 6. ELISA of antibodies

Equipment and reagents

- Microplate coated with peptide by any of the procedures described earlier, or with purified protein. *Protocol 1* is commonly used for proteins.
- Sample buffer (50 mM phosphate buffer, pH 7.4, 1.2% w/v NaCl, 0.05% w/v Tween 20, 1% w/v BSA).
- TMB peroxidase substrate. Dissolve 0.1 g of 3,3',5,5'-tetramethylbenzidine (Sigma T2885) in 0.1 ml of dimethyl sulphoxide (Sigma D5879). Add 9.9 ml of 0.1 M sodium acetate, pH 6.0, and, immediately before

use, adjust to 0.01% w/v hydrogen peroxide. A stock solution of hydrogen peroxide (30% w/v can be obtained from Sigma (H1009). Dynatech Laboratories market ready-to-use TMB peroxidase substrate solutions.
- Conjugate buffer (PBS, 1% w/v BSA, 0.5% Tween 20).
- PBST
- Peroxidase-labelled second antibodies appropriate to the application can be obtained from numerous suppliers including: Serotec, Dako, Sigma, and Amersham.

Method

1. Prepare the sample in which anti-peptide antibodies are to be measured in the sample buffer. The dilutions used will vary widely. If the serum is from a hyperimmunized animal the range from 10^{-3} to 10^{-6} may be suitable. If human serum, possibly containing antibodies to a virus peptide, is used the starting range may be 10^{-1} to 10^{-3}.

2. Add 100 μl of antibody dilution to duplicate wells on the plate. Include suitable positive and negative control samples.

3. Incubate at room temperature for 1–5 h. The time can be shortened if the plate is gently agitated. It is important to cover the plate and to avoid temperature gradients.

4. Wash the wells four times with PBST.[a]

5. Add a suitable peroxidase-labelled second antibody at an appropriate dilution in conjugate buffer.[b,c,d] Most manufacturers give recommendations for ELISA working dilutions; often a 10^{-3} dilution.

6. Cover the plate and incubate at room temperature for 30 min. Shaking is optional.

7. Thoroughly wash the wells with PBST; 10 times per well is not excessive at this stage.

8. Add 100 μl of TMB peroxidase substrate to each well.

147

Protocol 6. *Continued*

9. Cover the plate with aluminium foil and incubate, preferably on a microplate shaker, for 10–60 min at room temperature. A blue reaction product forms.

10. The reaction may be stopped if necessary by the addition of 50 μl of 6% (v/v) phosphoric acid to each well. This turns the reaction product yellow.

11. Read the unstopped reaction product at 650 nm or the stopped reaction product at 450 nm on a plate reader. The stopping procedure increases the sensitivity by approximately three-fold.

[a] Many workers use automated plate-washers. It is important that these are regularly cleaned, and that their correct performance is checked. Microbial growth can create havoc by blocking wash jets.

[b] A second antibody reagent referred to as, for example, anti-mouse IgG' will invariably contain antibodies to immunoglobulin light chains unless these have been specifically removed. Such a labelled second antibody would thus react with other isotypes of the same species, for example, IgM. Antibodies specific for individual isotypes would need to react only with heavy-chain epitopes. Suppliers of labelled second antibodies should be ask⁳d to supply detailed information on the specificity of the product before an order is placed.

[c] Antibodies to rat IgG will cross-react extensively with mouse IgG, and significantly with human IgG, unless the cross-reacting antibodies are adsorbed out. Several manufacturers, for example Amersham, sell species-specific second antibodies for ELISA. However, it is up to the user to determine whether the degree of species specificity it good enough for the intended application.

[d] There is no reason why this procedure should not be carried out using alkaline phosphatase-labelled second antibodies. Details of the use of such conjugates and the substrate are given in Section 3.4.3.

The simplicity and elegance of ELISA techniques conceal a large potential for misinterpretation of results. There are many possible reasons why the signal observed may not be due to the specific interaction required, or a signal not observed where it might be expected. These include impurities in the peptide or parent protein used for immunization or immobilization, lack of specificity in the second antibody, or problems of immobilizing peptides and proteins in an orientation allowing interaction with antibodies. When establishing a new ELISA procedure it is useful if control samples are tested. In many studies these would include sera expected to contain high levels of antibody alongside normal sera from the same species. Remember that because of the immense diversity of antibody specificities in serum all sera will react in ELISA with all antigens if tested undiluted. A useful confirmation that anti-peptide antibodies reacting with an immobilized parent protein are specific is the complete inhibition of the ELISA reaction by pre-incubating the serum dilution with an excess of either the immunizing peptide or the parent protein. Immunoblotting is a particularly valuable method of demonstrating the specific reaction of anti-peptide antibodies with the parent protein. This is discussed in Section 4.

2.4 Selection of monoclonal anti-peptide antibodies with the highest functional affinity for the parent protein

Following synthetic peptide immunizations it is necessary to select the best clones from those producing antibodies against the parent protein. Several points should be noted:

(a) Many proteins become partially denatured on binding to plastic (16). Thus it is possible that some antibodies selected by their strong ELISA reaction may not react so well with the same protein in solution.

(b) Coupling of peptides to carrier proteins ensures adequate recognition by helper T-lymphocytes, and has been observed to produce antibodies of higher average affinity (17).

(c) The affinity with which anti-peptide monoclonal antibodies react with the parent protein will be influenced by the accessibility of the epitope and the flexibility of the protein surface (18).

(d) The frequency of B-lymphocytes making high-affinity antibodies to the parent protein may be very small. ELISA screening of fusion super-natants should be rigorous. In the author's work on the preparation of monoclonal antibodies to the βA subunit of inhibin (5) only six positive wells were found on 20 microplates. Five of these were low-affinity IgM antibodies. The sixth was a high-affinity IgG, which has proved useful for immunoassay, immunocytochemistry, and immunoblotting of inhibins and activins (5, 6).

Rigorous procedures for measuring antibody affinity have been described elsewhere, but the majority of researchers making monoclonal antibodies need simpler procedures to rank their antibodies without delay. Two procedures useful for this purpose will be described.

2.4.1 Inhibition ELISA

In this procedure the various monoclonal antibody preparations are first titrated by ELISA on plates coated with the protein of interest, for example as in *Protocol 6*. From this initial titration the dilution of each antibody which gives an absorbance of 0.5 is determined. This dilution is then made up in assay buffer, and incubated overnight at 4°C in three aliquots. One aliquot is made 30 ng/ml, and another 300 ng/ml, with the protein of interest, and the third aliquot has no addition. The following day the ELISA titration is re-peated exactly as before. The antibodies showing the most inhibition are judged to be of the highest functional affinity for the parent molecule. The advantage of this approach is that it is selective for antibodies which react well with the parent protein in solution. The disadvantage is the need to pre-titrate each antibody before inhibition can be carried out.

2.4.2 Sensitivity to mild denaturants

It has been observed that if ELISA titrations are done in parallel for an antibody in the presence and absence of the mild denaturant diethylamine, then antibodies of high affinity show less perturbation than those of low affinity. Concentrations of diethylamine (Sigma D3256) in the range 5–50 mM have proved suitable (15, 19). A good starting point is to test firstly the sera from immunized mice to be used for fusion. A concentration of diethylamine should be chosen which gives slight, but measurable, alteration to the titration profile. The hope would be to subsequently identify, in the monoclonal antibodies made from these mice, those with similar properties to antibodies in the serum. In practice most monoclonals show much greater inhibition, indicating that they are of lower affinity than the best antibodies in a polyclonal antiserum.

2.5 Peptide–carrier conjugates in immunoassay

One method for detecting anti-peptide antibody responses by ELISA is to attach the same synthetic peptide as used for immunization with similar chemistry (and thus orientation) to a different carrier protein to that used for immunization and coat it on to plates for ELISA (*Protocol 1*). It is known that very often the immunodominant regions of the peptide are those distal to the point of attachment to the carrier (18). It would not be logical to immunize with a peptide coupled to a carrier protein via a cysteine at its C-terminal (biasing the antibody response to N-terminal) and then screen with the same peptide attached by its N-terminal to a different carrier by glutaraldehyde (biasing the detectability away from the N-terminal). One potential problem is that some of the linkage agents used to couple peptide to carriers may themselves give rise to antibodies making it difficult to distinguish antibodies to the peptide from antibodies to the link, even if the carrier is changed (20). Possible solutions to this problem include the use of a different heterobifunctional chemical to link the peptide to carrier for immunization and screening (a wide range of heterobifunctional agents is available from Pierce), or the use of weakly immunogenic linkers. It has been claimed that one heterobifunctional agent *N*-maleimido-6-aminocaproyl-(2'-nitro,4'sulphonic-acid)-phenyl ester (mal-sac-HNSA) is only very weakly immunogenic (20). *Protocol 7* shows the procedure used in the author's laboratory to couple thiol-containing peptides to carrier proteins using this reagent.

Protocol 7. Coupling of a thiol-containing peptide to carrier protein[a]

Reagents

- Carrier proteins: keyhole limpet haemocyanin (Pierce 77100), ovalbumin (Pierce 77120), or BSA (Pierce 77110) in protease-free form ready for use[b]
- Mal-sac-HNSA (Bachem Q-1615)
- Sephadex G25 and G100 (Pharmacia)

Method

1. Reconstitute the carrier protein in water to make a 10 mg/ml solution. This gives the protein in a neutral pH phosphate buffer. If the proteins are bought from other suppliers, 0.1 M phosphate buffer, pH 7.4, would be suitable.

2. To 1 ml of a 10 mg/ml solution of the chosen protein add 10 mg of mal-sac-HNSA. Mix and then stir the solution with a small magnetic stirring flea for 30 min at room temperature.

3. Desalt the sample on a 1 × 20 cm column of Sephadex G25 equilibrated with 0.1 M phosphate buffer, pH 6. Collect 1 ml fractions. (The carrier proteins elute after about 6 ml. Haemocyanin is easily visible by its slight opacity. Fractions containing ovalbumin or BSA will have a faint yellow colour but their location can be checked by reading a dilution of several fractions at 280 nm.)

4. Pool the fractions containing the activated carrier protein in as small a volume as possible.

5. Dissolve the thiol-containing peptide to be coupled in 0.1 M phosphate buffer, pH 7.5, and immediately add it to the activated carrier. Adjust the pH to 7.5 by adding 0.5 M phosphate buffer, pH 7.5, dropwise.[c]

6. Flush the tube with nitrogen and agitate it on a shaking platform overnight at room temperature.

7. Remove any free peptide from the peptide–carrier conjugate by chromatography on a 100 × 2 cm column of Sephadex G100 equilibrated with a buffer containing 50 mM phosphate buffer, pH 7.0, 1.2% (w/v) NaCl in the cold-room. Pump at 20 ml/h.

8. Collect the first peak. Check the concentration by protein determination and freeze the solution in aliquots.

[a] For the purpose of making peptide attached to irrelevant carriers for detecting antibody responses it is not usually necessary to check the coupling efficiency.

[b] It is important that the carrier proteins have low protease activity, and that they are free of substances which may interfere with coupling, for example Tris, azide.

[c] Some peptides precipitate out during coupling, particularly to haemocyanin.

2.6 Biotinylated peptides in immunoassay

A promising recent development is the idea of using peptides biotinylated during synthesis at either their N- or C-terminal and immobilized on streptavidin-coated microplates. Biotinylation of peptides on the resin during synthesis is straightforward (21, 22) and the biotin–peptide linkage is stable to most of the procedures used to detach peptides from the resin. The high affinity of the biotin–streptavidin interaction takes care of the immobilization and the peptide is held in a predetermined orientation.

Peptides from the N-terminal of a protein of interest held by a C-terminal biotin would be expected to share conformations with the N-terminal in the parent protein. Peptides from the C-terminal of a protein held by an N-terminal biotin should resemble the C-terminal of the parent protein. An advantage of this approach to peptide immobilization is that by saturating the available streptavidin sites many different peptides can be coupled to the plates in identical molar amounts. In other procedures the amounts of different peptides which bind can vary widely. Non-specific binding of the peptide to the plastic can be minimized by the presence of albumin and Tween in the solution. In other coating procedures aiming at covalent attachment of peptides a proportion of the bound peptide will be simply adsorbed. A problem is the loss of solubility which some peptides may undergo on bio-tinylation but it may be possible to dissolve the peptide in a solvent such as dimethylformamide or dimethyl sulphoxide before diluting it to the low concentration used for coating the streptavidin plates (100 ng/ml). It is quite possible that at this low concentration the peptide will stay in solution long enough to attach to streptavidin. One other application of these streptavidin-coated microplates is to immobilize biotinylated antibodies for two-site ELISAs (see Section 3.4.5)

Protocol 8. Attachment of biotinylated peptides to streptavidin-coated microplates[a]

Equipment and reagents

- Streptavidin-coated microplates or strips (Bio-Products or Labsystems) or ELISA-grade microplates and streptavidin (Sigma S4762)
- PBS (*Protocol 1*)
- BSA (*Protocol 1*)
- PBST (*Protocol 1*)
- Peptide buffer (50 mM phosphate buffer, pH 7.0, 1% w/v BSA, 1.2% NaCl, 0.5% Tween 20)

Method

1. If streptavidin-coated plates are not available coat the wells of a standard microplate by overnight incubation at room temperature with 100 μl of streptavidin (5 μg/ml) in 50 mM carbonate buffer, pH 9.6[b]

2. Shake the plate dry and add 150 μl of PBS, 1% (w/v) BSA. Allow it to stand for 1 h at room temperature.

3. Wash the plate thoroughly with PBST.

4. Shake the plate dry and add 100 μl of biotinylated peptide (0.1 μg/ml) dissolved in peptide buffer to each well. Incubate at 4 °C overnight.

5. Wash the plate thoroughly with PBST and use it immediately for ELISA (*Protocol 6*).

[a] Procedures for adding biotin to the N-terminal (21) and the C-terminal (22) of a peptide have been described.

[b] Streptavidin does not adsorb very well to ELISA microplates so that, if a high-density, stable coat is required, commercially prepared plates in which the attachment is covalent are to be preferred.

2.7 Applications of branched peptides

Recently it has been shown that branched peptides built from an octameric core of lysines are extremely good in solid-phase ELISA procedures for the detection of antibodies (23). The branched peptides functioned best when a spacer of four to five glycine residues was introduced into the structure before the sequence of interest (23). It was suggested that the reasons these branched peptides performed so well were: their superior adsorption characteristics, their greater reactivity with antibodies in the adsorbed state, and their participation in multivalent binding which enabled the detection of lower-affinity antibodies (see Chapter 4).

2.8 The importance of peptide purity

All synthetic peptides will contain, as contaminants, sequences other than the target sequence, and also sequences containing small amounts of modified or still protected side-chains. If the same synthetic peptide preparation is used for both immunization and screening it is possible that highly immunogenic impurities may be detected preferentially. Polyclonal antibodies made using such peptides, or even affinity purified using them, may give undesired cross-reactions. The likelihood of this situation occurring is reduced if peptides used for immunization and screening are purified to homogeneity on HPLC. By contrast, in the author's experience, if the parent proteins are to be used for screening fusions for preparation of monoclonal antibodies, purification of peptides is usually unnecessary. In this case the presence of antibodies to error sequences etc. is simply ignored by the screening assay.

One strategy the author has found helpful is to use peptides made in his laboratory by Fmoc chemistry for immunizations, but similar commercially obtained sequences made using Merrifield Boc chemistry for the screening. Any ELISA reactivity then observed is most likely due to genuine reactions with the correct sequence common to both.

3. Immunoassays for the parent protein using anti-peptide antibodies

3.1 Development of immunoassays for molecules unavailable for antibody production

Antibodies raised to synthetic peptides can be used in many of the same immunoassay procedures as those made by conventional methods, and this is not the place for a comprehensive treatment of immunoassay methodology. Instead, emphasis will be placed upon a unique application of anti-peptide antibodies, to construct sensitive and specific immunoassays for the products of newly sequenced genes, without pure material being available for immunization.

The following issues will be addressed:

- How have anti-peptide antibodies been used in immunoassays?
- What factors have to be taken into account in using anti-peptide antibodies for immunoassay?
- What are the limitations on the use of anti-peptide antibodies in immuno-assays?

Protocols successfully used in the author's laboratory will be described and examples of the results given.

Two categories of immunoassay will be considered:

(a) Competitive assays. The analyte in the sample is made to compete with labelled standard analyte for occupancy of a limited amount of specific antibody. In such assays the sensitivity of detection is limited by the affinity of the antibody and the precision with which the occupancy of the antibody binding sites can be measured experimentally.

(b) Two-site (sandwich) assays. In this case the analyte must have two different epitopes (or a repeating epitope). One antibody is attached to a solid phase (or has a label to permit this to take place), and the other antibody either directly carries a radioisotopic, enzymic, fluorescent, or chemiluminescent label or a means of subsequently binding one. In such assays the sensitivity of detection is limited largely by the size of the non-specific signal from a zero-analyte sample compared with that from the specific signal from small amounts of analyte (24, 25).

Although two-site assays can often be made more sensitive than competitive assays using the individual antibodies alone the sensitivity limit of all immunoassays is more likely to be due to the affinity of the antibodies than the detectability of the label.

3.2 Purification of antibodies for use in immunoassays

The degree of antibody purity needed depends on the type of assay chosen:

- Monoclonal antibodies and polyclonal antibodies for competitive assays can be used unpurified.
- Monoclonal antibodies for two-site assays need to be purified to an IgG fraction. A very effective method of achieving this is with immobilized Protein A. Procedures for the use of Protein A are described elsewhere (26). (See Chapter 5.)
- Polyclonal anti-peptide antibodies for use in two-site assays are best affinity purified on a column containing peptide (*Protocol 9*). (See also Chapter 5, *Protocol 1*.)

Protocol 9. Immobilization of a thiol-containing peptide and its use in affinity purification of anti-peptide antibodies[a]

Reagents

- 3-maleimidobenzoyl-N-hydroxy-succinamide (MBS) ester (Pierce 22312)
- PBS (*Protocol 1*)
- Sepharose-AH (Pharmacia)
- Dimethylformamide (DMF)

A. *Immobilization of peptide*

1. Wash Sepharose-AH thoroughly on a fine sinter with PBS.

2. Mix 3 ml of packed gel with 6 ml of PBS.

3. Add 500 µl of MBS (5 mg/ml in DMF).

4. Mix end-over-end for 2 h.

5. Wash the Sepharose on the sinter with PBS and then remove it to a tube containing 4 ml of PBS.

6. Add 5 mg of thiol-containing peptide.[b]

7. Mix end-over-end at room temperature overnight.

8. Add an equal volume of 1 M Tris–HCl buffer, pH 7.4, and mix for 1 h.

9. Wash the immunoadsorbent exhaustively with 0.2 glycine–HCl buffer, pH 2.6.

B. *Immunoaffinity chromatography*

1. Pack a column (0.8 × 4 cm) with the gel and equilibrate with PBS.

2. Dilute the serum which contains anti-peptide antibodies two-fold with PBS and centrifuge it at $100\,000\,g$ for 30 min.

3. Pump on to the column at about 1–2 ml/h in a cold-room.

4. Wash the column with about 100 ml of PBS until the absorbance at 280 nm is less than 0.05 (1 cm cell).

5. Elute the specific antibodies with 0.1M glycine–HCl buffer, pH 2.6.

6. Collect 2 ml fractions into tubes containing 1 ml of 1 M Tris–HCl buffer, pH 7.4.

7. Dialyse away the Tris completely by three changes to 2 litres of PBS each time.

8. Use the antibody immediately or store frozen in aliquots.

---

[a] The procedure is essentially that described by Chersi *et al.* (27).
[b] This is the amount suitable for a 15-mer peptide.

3.3 Competitive assays

Two examples of competitive assays using antibodies raised to synthetic peptides in RIAs will be described (3, 28). In both cases it was possible to establish assays applicable to biological fluids. The idea behind each assay was the same. Synthetic peptides were used both as radiolabelled tracer, and as immunogen to produce the necessary antibodies, and these were used in a conventional RIA format.

There are several advantages in this approach:

(a) It can be applied to assay a molecule of interest immediately the N-terminal sequence becomes available, either from nucleic acid sequencing or the microsequencing of small amounts of purified material.

(b) The antibodies used can be polyclonal (e.g. rabbit or sheep) and can be used for other purposes such as immunoblotting and immunocytochemistry.

(c) It is commonly the case that purification of many proteins of interest from natural sources or recombinant expression systems requires a tedious bioassay. The availability of a simple and specific immunoassay for the molecule of interest can often replace this for many purposes.

In principle this type of assay could be applied to many molecules where the N- or C-terminal is exposed, and where antibodies to peptides from N- or C-terminal sequences give rise to antibodies which react strongly with the parent protein. A few points to bear in mind about this approach are as follows:

(a) A portion of the anti-peptide antibodies may react with the tracer, but not with the parent protein. This will mean that the tracer cannot be completely displaced down to the non-specific binding level by the analyte or fluids containing it. Although 10 or 20% non-inhibitable binding could still allow a usable standard curve to be generated, at some point the assay may be judged unsatisfactory.

(b) Purification of peptides for immunization, and particularly for radio-labelling, is essential.

(c) Care must be taken to show that any displacement of radiolabel by tissue extracts or body fluids is not due to contaminating proteases degrading the peptide tracer (29).

(d) A wide range of different conditions for RIA are possible and the literature on RIA is extensive (26, 30, 31). Potential users should consider the safety implications of work with radioiodine. *Protocol 10* describes the establishment of an RIA using anti-peptide antibodies and radiolabelled peptide.

(e) It is preferable that the peptide sequence used for iodination contains only one tyrosine, and that this is not at the immunodominant end of the peptide, i.e. at the end opposite to that used for carrier attachment. The peptide used for radiolabelling may be shortened to $1-10[\text{Tyr}^{11})]$ if necessary to achieve this.

(f) If difficulties are encountered using the N-terminal sequences to establish assays of this type, a similar approach could be tried at the C-terminal using peptides with tyrosine at their N-terminal for use as tracer and with cysteine for carrier conjugation and immunization.

A procedure for establishing an RIA is shown in *Protocol 10*.

Protocol 10. Radioimmunoassay

Reagents

- Antiserum (e.g. rabbit) prepared by immunization with the 1–20 or 1–30 sequence of the protein of interest coupled to a carrier protein (*Protocol 7*) by a cysteine added to the peptide at its C-terminal during synthesis ($1-20[\text{Cys}^{21}]$ or $1-30[\text{Cys}^{31}]$). If the longer peptide is hydrophobic and insoluble use the other.
- Purified synthetic peptide corresponding to $1-10[\text{Tyr}^{11}]$ or $1-20[\text{Tyr}^{21}]$ for radiolabelling. Procedures for radioiodination with ^{125}I are well established (30, 31) and will not be described here. Iodinated peptide tracers should have a specific activity of 30–100 μCi/μg, and may be usable for up to 6 weeks.

- RIA diluent: PBS, 1% (w/v) protease-free BSA (Sigma A3294). For some assays non-specific binding may be lowered by addition of 0.1% (w/v) Tween 20 or Triton X100.
- Solid-phase second antibody, SaCell (Immunodiagnostic Services). Anti-rabbit and anti-sheep would be the commonest for polyclonal antibodies. This material contains second antibody on a microcrystalline cellulose.
- LP3 polystyrene tubes (Luckham).
- Gamma counter

A. *Antibody titration*

It is necessary to establish whether the antiserum can bind the tracer peptide.

1. Make a serial doubling dilution of the antiserum in RIA diluent from 10^{-3} to 10^{-6}.
2. Add 100 μl of each dilution to LP3 tubes. Also add 100 μl of diluent to two tubes as a test for non-specific binding. Add a further 100 μl of RIA diluent to each tube in the place of the sample or standard which will be present in the final assay.
3. Add 50 μl (20 000 c.p.m) of radiolabelled peptide tracer in RIA diluent to each tube. Also add 50 μl to two further tubes as 'totals'.
4. Incubate the tubes overnight at 4 °C.
5. Add 50 μl of undiluted SaCell suspension to each tube, except the 'totals'.
6. Vortex each tube and allow to stand for 30 min.
7. Add 2 ml of water to each tube and then centrifuge at 5000g for 15 min.

Protocol 10. *Continued*

8. Decant the supernatant and discard. Blot the tube on paper towelling to remove the last traces of liquid.

9. Count the tube in a gamma counter.

10. Note the maximum percentage of the total counts which are bound. With a fresh tracer this should be 70–90% but this will decline as the tracer ages.

11. Estimate the antibody dilution needed to bind approximately 30% of the maximum which is bound at low dilution. If the primary antiserum is of very low titre it may only be possible to estimate the dilution which binds 30% of the total counts.

B. *Assay procedure*

1. Make up a set of standard concentrations of the analyte in the RIA diluent. If no standard is available use serial dilutions from a tissue fluid or extract known to contain significant amounts of the analyte.

2. Set up 'zero analyte' tubes in duplicate as follows: 100 μl of antiserum dilution as determined earlier plus 100 μl of diluent.

3. Set up the standards and samples as follows: 100 μl of antiserum plus 100 μl of standard or dilution of samples.

4. Set up non-specific binding tubes in duplicate containing 200 μl of RIA diluent.

5. Incubate the tubes overnight at 4 °C. This can be shortened to a few hours if reduced sensitivity is acceptable.

6. Add 50 μl of the tracer to each tube and to the duplicate 'totals'.

7. Incubate overnight at 4 °C.

8. Repeat the SaCell separation (see above).

9. Count the tubes in a gamma counter. Most instruments will carry out data reduction and fit a standard curve.

10. If the assay is working there should be near complete displacement of the tracer by excess of both purified analyte and crude samples containing it.

The competitive form of assay works best if the analyte is present at sufficiently high concentrations in biological fluids or tissue extract to allow dilution. It may sometimes be necessary to add protease inhibitors (29).

Similarity between the purified analyte and analyte-related immuno-reactive material in samples can be established by plotting the RIA data on logit/log graph paper (Heffer's) and examining the plots for linearity and parallelism. A marked non-parallel response may indicate that the analyte in the standard and the analyte in the body fluid are not reacting with the antibody in identical fashion. Accurate estimates of the relative amounts of analyte in unknown samples can only be obtained if the unknown results are interpolated from a standard curve with which both serial dilutions of unknowns and standards conform. It is possible to use dilutions of crude biological fluids as standards for the purpose of immunoassay, assigning the undiluted fluid a concentration in arbitrary units. This assay format may allow sensitive assays to be established for molecules whose genes have just been cloned, and before purified material is available. In this case dilutions of a biological fluid known to contain the material may be the only available standard.

Two examples of RIAs for peptides will be described:

(a) RIA for parathymosin (3). In this work polyclonal antibodies were raised to rat parathymosin by immunization with the 1–30 N-terminal sequence coupled to haemocyanin. A 1–12[Tyr^{13}] peptide was radioiodinated and used as the tracer. In a conventional RIA procedure the radiolabel could be completely displaced by excess parathymosin α, and a standard curve used for estimation of the amount of parathymosin in various samples. Serial dilutions of crude rat liver extract gave an RIA dose–response curve superimposable on (parallel to) that obtained with parathymosin standard, and the partly homologous molecules thymosin α and prothymosin α did not cross-react.

(b) RIA for inhibin. In this work polyclonal antibodies raised to N-terminal peptides of the α subunit of 32 kDa human ovarian inhibin were used to establish an RIA for inhibin-like materials in various fluids. The tracer used was either [Tyr^0]inhibin α(1–32) (28) or inhibin α(1–25)[Gly^{26}Tyr^{27}] (29). An example of an RIA for inhibin-like immunoreactivity in various fluids is shown in *Figure 1*.

Conserved molecules may be poor immunogens and give rise to low-affinity antibodies. This may limit the sensitivity of detection in RIA. For such molecules, even if they are available in large amounts, there may still be advantages in using N- or C-terminal peptide sequences for immunization to produce antisera, but replacing the peptide tracer in RIA with iodinated recombinant material. Coupling of the peptide to a carrier protein breaks self-tolerance to give higher-affinity antibodies (17). There is no doubt that antibodies have been made using synthetic peptides where conventional approaches had been unsuccessful (5, 6).

3.4 Two-site (sandwich) assays

3.4.1 Are two-site immunoassays more specific?

Since in a two-site assay the analyte is required to react with two different antibodies it might be thought that such assays would always be less prone to

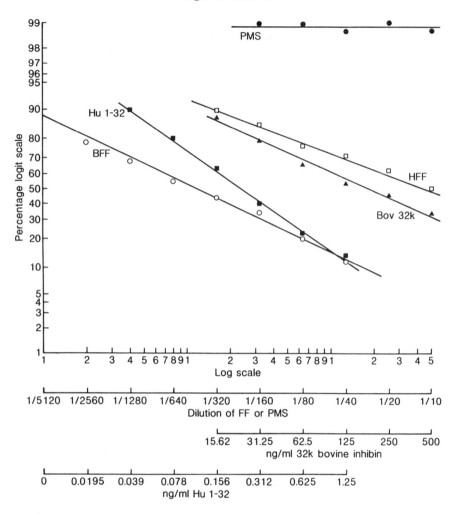

Figure 1. Radioimmunoassay of inhibin-like immunoreactivity in biological fluids. The assay used [Tyr⁰] inhibin α(1–32) and a sheep antiserum prepared by immunization with inhibin α(1–29) [Tyr³⁰] attached to ovalbumin with bis-diazotized toluidine. The data show that the dose response of bovine follicular fluid (BFF) diluted in parallel with the purified 32 kDa bovine inhibin standard (Bov 32 k). The dose response with human follicular fluid (HFF) was almost parallel to the 32 kDa standard, but human inhibin peptide α 1–32 (Hu 1–32) was not parallel with the response of these fluids or the 32 kDa standard. Post-menopausal human serum (PMS) had no detectable immunoreactivity. The RIA procedure was essentially as in *Protocol 10*. (Reprinted from (28) by kind permission of the publishers, Mary Ann Liebert, New York.)

cross-reactions with related or unrelated molecules in samples. By contrast, it has been shown theoretically and practically (32) that on occasions two-site assays can give a high degree of cross-reactivity where it would not have been expected. Two-site assays use excess antibodies and this will serve to increase the apparent cross-reactivity of molecules with limited structural similarity to the analyte of interest. Boscato *et al.* (32) describe a two-site assay for human chorionic gonadotrophin in which a strong cross-reaction was observed with the β subunit, even though in solution-phase RIAs one of the two antibodies had negligible reactivity with the β subunit. In the author's laboratory a two-site assay developed for inhibin using monoclonal antibodies to the α and β A subunits of inhibin recognized inhibin (α/βA), but also detected activin (βA/βA) with a 5% cross-reaction (6). We recently found that in a reconfigured version of this assay for inhibin, with the antibodies used for labelling and capture exchanged, the cross-reaction of activin was only 0.05% (Groome and O'Brien, unpublished). This example should serve to demonstrate that there is much that is empirical about establishing optimized immunoassays. Once the general principles are understood the antibodies have the final word in the specificity and sensitivity of assay they can deliver.

3.4.2 Are anti-peptide antibodies generally more cross-reactive than antibodies raised to parent proteins?

Antibodies raised to synthetic peptide immunogens often appear to recognize continuous epitopes as short as four to five amino acids if these are mapped with the Pepscan method described by Geysen (33). Fortunately, it is seldom necessary to define the epitope this precisely in order to use an antibody for immunoassay but this finding does have implications for the specificity of such antibodies. Such short sequences may be shared with proteins unrelated to the protein of interest. Since antibodies are used in excess in a number of immunological procedures such a immunoblotting, immunocytochemistry, and two-site assays, there is a strong possibility that the antibodies will detect molecules other than those expected. A striking example of an anti-peptide monoclonal antibody cross-reacting with numerous other molecules is given by Khachigian *et al.* (34). This antibody was observed to react with platelet-derived growth factor, polylysine, myoglobin, and BSA (34). It was established that the amino acid sequence LysLys was sufficient to constitute an epitope, and this could be presented in a variety of different conformational frameworks.

The possibility that both monoclonal and polyclonal anti-peptide antibodies may react with molecules other than those containing the sequences used as immunogen to produce them must be borne in mind by researchers using such antibodies. Antibodies, however made, must be proved experimentally to be sufficient for the purpose for which they are needed. This is an important factor when many research groups frequently use antibodies made by others.

3.4.3 Setting up a two-site ELISA with the minimum of reagent preparation

Establishing and optimizing immunoassays can be a time-consuming and tedious process. In the author's experience the following procedures allow a very rapid assessment to be made of how well a pair of monoclonal antibodies perform in two-site assays. Initially for this approach it is necessary to bio-tinylate the antibodies. This is a straightforward procedure carried out as in *Protocol 11*. *Protocol 12* shows how to test the antibodies in a two-site ELISA assay with a standard peroxidase assay for detection. *Protocol 13* uses the recently described technique of catalysed reporter deposition (35) to amplify the signal from peroxidase, and can often result in enhanced sensitivity of an assay. A similar assay can be established using alkaline phosphatase as the enzyme label; *Protocol 14* is based on a recent modification (36) of a pub-lished procedure for amplifying the alkaline phosphatase assay (37). Finally *Protocol 15* describes a very sensitive assay procedure based on the use of streptavidin–galactosidase with a fluorogenic substrate.

It should be noted that only antibodies of high affinity will permit the development of immunoassays detecting pg/ml amounts of analyte. To obtain such monoclonal antibodies it is advantageous if small amounts of the parent protein are available, even in crude form, for use in ranking the affinities of various candidate antibodies by procedures such as those mentioned in Section 2.4.

In the present chapter only the first stage of immunoassay development (i.e. assays which work with high sensitivity in simple buffers) will be discussed. Application of assays to serum and plasma can pose many difficulties. Prob-lems of matrix interferences have been thoroughly discussed by Wood (38). Heterophilic antibodies, which can cause false-positive reactions, have often been successfully overcome by adding up to 5% v/v normal mouse serum to the assay diluents or, in extreme cases, by the use of F(ab')$_2$ or Fab fragments of one of the antibodies (39). Methods for preparing antibody fragments and for directly coupling them to enzymes for use in two-site assays have been reviewed by Ishikawa *et al.* (40).

Protocol 11. Biotinylation of antibodies

Reagents

- Sulphosuccinimidyl-6-(biotinamido) hexa-noate (NHS-LC-Biotin) (Pierce 21335)
- Dimethyl sulphoxide (DMSO) (Sigma D5879)
- PBS

Method

1. Dialyse the purified antibody overnight against several changes of 0.1 M sodium bicarbonate. The pH is around 8.1.

---

2. Adjust the antibody concentration to 1 mg/ml and place the solution in a container containing a magnetic stirring bar. A 1 mg/ml solution of purified IgG has an absorbance at 280 nm of 1.4 in a 1 cm cell.

3. Make up a solution of 2 mg/ml NHS-LC-biotin in DMSO.

4. Add a total of 50 µl of this solution to each ml of antibody in ten 5 µl aliquots over a minute.

5. Leave the solution stirring for 4 h at room temperature.

6. Dialyse away the uncoupled biotin against PBS for two days in a cold-room. Change the buffer twice each day.

7. Store the biotinylated antibody at 4 °C in the presence of 1% (w/v) BSA and 0.1% (w/v) NaN_3.

Protocol 12. Two-site (sandwich) ELISA for protein antigens[a, b]

Equipment and reagents

- PBST (*Protocol 1*)
- PBST-BSA (PBST, 1% (w/v) BSA)
- Purified IgG of each of two monoclonal antibodies
- PBS

- Biotinylated IgG of each of two monoclonal antibodies (*Protocol 11*)
- Streptavidin–peroxidase conjugate (Gibco-BRL 560-9534SA)
- ELISA-grade microplates (Nunc 4-39454A)

Method

1. Coat the wells of half of the plate overnight at 4 °C with 100 µl of each purified antibody (10 µg/ml) in 50 mM sodium carbonate/bicarbonate buffer, pH 9.6.[c]

2. Shake the plate dry and add 150 µl of PBS, 1% (w/v) BSA to block excess adsorption sites. Stand the plate at room temperature for 1 h.

3. Wash the plate thoroughly with PBST. Allow it to stand for 1 h in this solution to remove loosely bound antibody.

4. Make dilutions of standards or unknowns containing the analyte of interest in PBST-BSA.

5. Shake the plate dry and add triplicate samples (100 µl) to wells on the plate. (It is common practice to avoid the edges of microplates, which may give variable results.) Include wells with diluent only. Add one set of standards to each half of the plate coated with a different antibody.

6. Cover the plate and gently shake it for 2–5 h at room temperature. This stage can be done in a cold-room if the analyte is more stable. Overnight incubation may be needed for some anti-peptide anti-bodies with a slow association rate for the reaction with the parent protein.

163

Protocol 12. *Continued*

7. Wash the plate with PBST and add 100 μl of each biotinylated second monoclonal antibody to the appropriate half of the plate. A suitable concentration should be 1–10 μg/ml in PBST-BSA. Incubate for 1–2 h at room temperature.

8. Wash the plate with PBST very thoroughly, shake dry, and add 100 μl of streptavidin–peroxidase diluted ($\times 10^3$) in PBST-BSA.[d]

9. Incubate the plate on a shaker for 30 min at room temperature.

10. Wash the plate thoroughly (10 times) with PBST and shake it dry on paper towelling.

11. Add 100 μl of the TMB peroxidase substrate to each well as in *Protocol 6*.[d]

12. Stop the reaction when required with 50 μl of 6% (v/v) phosphoric acid and read the absorbance of each well at 450 nm on an automated plate reader. Most plate readers will perform standard curve-fitting and dose interpolation for unknowns. If additional sensitivity is needed then try catalysed reporter deposition (*Protocol 13*).

[a] Optimization of two-site assays can require a lot of work, but the above protocol is likely to give reasonable results if the particular antibody pair has much assay potential.

[b] The procedure described will reveal which antibody is best for capture and which for use in the fluid phase. Some antibodies may lose much of their binding affinity when adsorbed to a surface. In the author's experience an orientated coupling of the antibody to hydrazide plates (AvidPlate-HZ) through the carbohydrate residues in the Fc gives a stable and reproducible solid phase (*Protocol 16*).

[c] The occasional monoclonal antibody may adsorb best at some other pH, but the buffer described is widely used for both monoclonal and affinity-purified antibodies.

[d] There is no reason why this assay should not, at step **8**, use streptavidin–alkaline phosphatase and, at step **11**, alkaline phosphatase substrate. It is recommended that the wash solution is 50 mM Tris–HCl buffer, pH 7.5, 0.05% Tween 20 because alkaline phosphatase is inhibited by phosphate buffers. This solution, supplemented with 1% (w/v) BSA, is used to dilute the streptavidin–alkaline phosphatase conjugate. The substrate is 5 mM *p*-nitrophenyl phosphate (Sigma N2507) in 0.9 M diethanolamine buffer, pH 9.8, 0.5 mM MgSO$_4$. The reaction is stopped by the addition of 100 μl of 0.1 M EDTA and the yellow colour is read at 405 nm. Details of an amplified alkaline phosphatase assay are given in *Protocol 14*.

Three sensitive methods for the measurement of enzyme labels are described. The first is a way of amplifying the colour generated in peroxidase-based ELISAs and the second achieves a similar benefit for alkaline phosphatase systems. The third makes use of the greater sensitivity with which 4-methyl-umbelliferone can be detected fluorimetrically in β-galactosidase assay. The author's laboratory has found the β-galactosidase system to be the easiest way to set up an initial assay with a new pair of antibodies and the alkaline phosphatase to be the more sensitive of the two colorimetric systems. It should be noted that using these highly sensitive label detection systems will not improve the sensitivity of analyte detection unless the antibodies used in the assay are of suitable affinity and specificity and if other aspects of the assay such as incubation times, antibody concentrations, diluent and wash solution compositions, and wash times are optimized.

Protocol 13. Amplification of the peroxidase assay[a]

Reagents

- Streptavidin–peroxidase (Gibco-BRL 560–9534)
- Biotin-*N*-hydroxysuccinimide ester (Sigma H1759)
- Tyramine (Sigma T7255)
- DMSO (Sigma D5879)
- PBST (*Protocol 1*)
- PBST-BSA (*Protocol 12*)
- Hydrogen peroxide (30% solution available as Sigma H1009)
- TMB peroxidase substrate (*Protocol 6*)

Method

1. Make biotinyltyramine (BT) as follows. Mix 40 mg of tyramine and 100 mg of biotin-*N*-hydroxysuccinimide ester in 1 ml of DMSO and incubate overnight with stirring. The solution is used directly and is assumed to contain 55 mg/ml of BT.

2. Carry out the unamplified peroxidase-based ELISA (*Protocol 12*)[b] up to step **10** but instead of adding substrate add 100 μl of 0.1 M borate buffer, pH 8.5, BT (10 μg/ml) solution[c] and hydrogen peroxide to 0.01%.

3. Incubate the plate without shaking for 15–30 min.

4. Wash the plate thoroughly[d] with PBST and add 100 μl of streptavidin–peroxidase (1/5000)[e] in PBST-BSA.

5. Incubate the plate without shaking for 30 min.

6. Wash the plate thoroughly with PBST and add 100 μl of peroxidase substrate to each well as in the unamplified assay (*Protocol 6*).

[a] This procedure is that described by Bobrow *et al.* (35). It is believed to work because peroxidase generates free radicals on the BT which then react with the protein on the surface of the microplate.

[b] When the author first tried out the amplified and non-amplified methods in parallel on the same plate the results of the amplified assay were very variable. This difficulty was overcome by passing the BSA solution through a 0.45 μm filter before use.

[c] A kit (ELAST™) to carry out the amplification reaction is available commercially from Dupont and contains reagents sufficient for 10 plates.

[d] Particular care should be taken to ensure that the wash solution used in automated plate-washers is freshly made and that all washings are extremely thorough.

[e] The concentrations of BT and streptavidin–peroxidase may need optimization for each individual assay, and it is important to note that no attempt should be made to amplify an assay until the non-specific binding in the unamplified assay has been minimized. Colour everywhere is the commonest first experience of most workers using amplified assays.

Johannsson *et al.* (37) described the use of an amplification system to permit more sensitive detection of alkaline phosphatase in immunoassay procedures. It involves the addition to the wells of a solution containing NADP which the alkaline phosphatase converts to NAD. This then functions as a coenzyme in a redox cycle involving diaphorase, alcohol dehydrogenase, ethanol, and *p*-iodonitrotetrazolium violet. The final colour development is several orders of magnitude greater in the same time period than with the unamplified assay. The use of semicarbazide to trap acetaldehyde is based on the paper by Brooks *et al.* (36) who showed that it increases sensitivity.

Protocol 14. Amplification of the alkaline phosphatase assay[a,b,c,d]

Reagents

- Diethanolamine (Sigma D8885)
- NADP (Boehringer-Mannheim 1179969). This is a critical reagent and must be very low in NAD to give low background signals
- Semicarbazide–HCl (Sigma S4125)
- Diethanolamine buffer-NADP (50 mM diethanolamine buffer, pH 9.8, 0.5 mM MgSO$_4$, NADP (240 µg/ml), 10 mM semicarbazide–HCl)
- Absolute ethanol (BDH 101076H)
- *p*-iodonitrotetrazolium violet (Sigma I8377)

- Alcohol dehydrogenase (Sigma A3263). This must be first dissolved in PBS, 1% BSA (see *Protocol 1*) and dialysed exhaustively in a cold-room to remove NAD. It can then be stored in frozen aliquots. [e]
- Diaphorase (Sigma D2381). This must be treated in the same way as the alcohol dehydrogenase. [e]
- Amplifier solution (25 mM phosphate buffer, pH 7.0, ethanol (40 µl/ml), *p*-iodotetrazolium violet (300 µg/ml), alcohol dehydrogenase (150 µg/ml), diaphorase (150 µg/ml))

Method

1. Carry out the unamplified ELISA with streptavidin–alkaline phosphatase[f] (*Protocol 12*) up to step **10**.
2. Add 80 µl of diethanolamine-NADP buffer to each well.
3. Incubate for 15–30 min.
4. Add 200 µl of amplifier solution to each well.
5. Incubate for 30 min.
6. Read the absorbance at 492 nm.

   [a] This procedure is based on that described by Brooks *et al.* (36). Additional practical and theoretical aspects of this amplification are described elsewhere (41).
   [b] A kit to carry out this amplification is available commercially from Dako.
   [c] As with the peroxidase amplification procedure (*Protocol 13*), washing must be very thorough and wash solutions freshly prepared if meaningful results are to be obtained.
   [d] No attempt should be made to use amplification until the ordinary assay is fully optimized and the zero-analyte wells have less than 0.01 absorbance.
   [e] The essence of getting this technique to work with in-house reagents is to remove free NAD from the diaphorase and alcohol dehydrogenase by dialysis and to use very high quality NADP.
   [f] Streptavidin–alkaline phosphatase conjugates made using heterobifunctional agents often give lower non-specific binding than those made using glutaraldehyde. Dako and Serotec conjugates are made using these agents.

An example of the use of the amplified alkaline phosphatase assay is the author's recent work on inhibin (42) in which as little as 2 pg/ml of the dimeric inhibin could be detected in an immunoassay where both of the monoclonal antibodies used had been raised to synthetic peptides.

Protocol 15. Fluorometric assay of β-galactosidase

Reagents

- Streptavidin–β-D-galactosidase conjugate (Gibco-BRL)
- PBST-BSA (*Protocol 12*)
- PBST (*Protocol 1*)

- 4-Methyl-umbelliferyl β-D-galactoside (Sigma M1633)
- Substrate buffer (10 mM phosphate buffer, pH 7.0, 0.15 M NaCl, 0.1% (w/v) BSA, 1 mM $MgCl_2$).

Method

1. Carry out the two-site ELISA (*Protocol 12*) up to step **8** and then replace the streptavidin–peroxidase with streptavidin–galactosidase (1/1000) in PBST-BSA.

2. Incubate for 1 h at room temperature.

3. Wash the plate very thoroughly (10 times) with PBST and add 100 μl of substrate buffer containing 4-methyl-umbelliferyl-β-D-galactoside (0.1 mg/ml) to each well.

4. Incubate at room temperature for 2–10 h and stop the reaction by the addition of 50 μl of 10% (w/v) $NaCO_3$.[a]

5. Read on a fluorescence microplate reader using an excitation wavelength of 355 nm and an emission wavelength of 460 nm.[b]

[a] The progress of the reaction can be visually assessed by viewing the plate on a UV transilluminator of the sort used to look at DNA bands. **Be sure to wear suitable eye protection**.
[b] Although solid black ELISA plates specially for fluorescence readers are available we obtain acceptable results with ordinary Nunc plates when using a Flow Laboratories fluorescence microplate reader.

Figures 2 and *3* show the results of using *Protocol 15* with two anti-peptide monoclonal antibodies to inhibin subunits made in the author's laboratory (6). The first occasion on which the assay was run allowed very sensitive detection of inhibin, and we recommend the fluorimetric system as a rapid way of establishing the sensitivity potential of an assay. Subsequently we adapted the assay to a colorimetric end point using the amplified alkaline phosphatase system (*Protocol 14*), with similar sensitivity. The results of this assay are shown in *Figure 4*, and the details are discussed in Section 3.4.5.

3.4.4 Simplification of two-site ELISAs for routine use

The use of biotinylated detector antibodies was suggested in the previous section since biotinylation is one of the easiest procedures for labelling

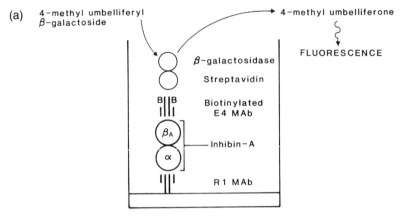

Schematic diagram of two−site assay for inhibin−A

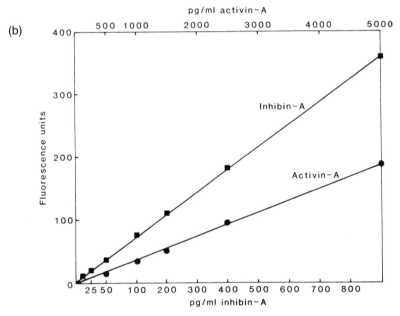

Dose-response relationship for two−site inhibin−A
assay with inhibin−A and activin−A

Figure 2. A two-site assay for 32 kDa recombinant inhibin using anti-peptide antibodies. (a) The diagram shows the capture of the inhibin by the immobilized R1 monoclonal antibody (to the α subunit), and detection by biotinylated E4 monoclonal antibody (to the βA subunit) followed by streptavidin−galactosidase and fluorogenic substrate. (b) Dose−response plots for inhibin-A and activin-A. Activin, a βA subunit dimer, showed a 5.2% cross-reaction in this assay. The assay procedure was essentially as in *Protocol 15*. (Reprinted from (6) by kind permission of the publishers, Elsevier.)

(a)

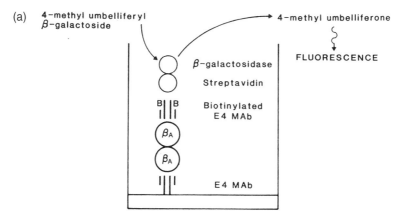

Schematic diagram of two−site assay for activin−A

(b)

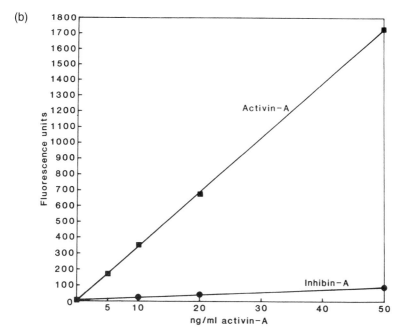

Dose-response relationship for two−site activin
assay with activin−A and inhibin−A

Figure 3. A two-site assay for activin using anti-peptide antibody. (a) The diagram shows the capture of activin by the immobilized E4 monoclonal and detection using biotinylated E4 followed by streptavidin–galactosidase and a fluorogenic substrate. This assay is possible because activin-A is a homodimer. (b) Dose–response plots for activin-A and inhibin-A. Inhibin demonstrates a significant cross-reaction of 5.3% in this assay. The assay procedure was essentially as in *Protocol 15*. (Reprinted from (6) with kind permission of the publishers, Elsevier.)

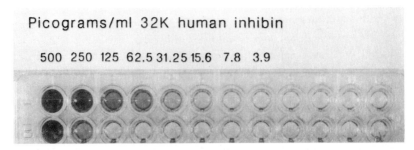

Figure 4. An amplified two-site assay for inhibin. The photograph shows in the top row the results obtained for various concentrations of 32 kDa inhibin standard, and in the second row the results obtained with dilutions of human follicular fluid starting at 10^{-3}) on the left and reducing in 10-fold steps. This reconfigured assay used the same anti-bodies (R1 and E4) as in our earlier assay shown in *Figure 2*. The biotinylated E4 monoclonal antibody was immobilized on a streptavidin-coated microplate (see text) and used to capture inhibin from the sample by its βA subunit. After overnight capture on a shaker at 4 °C the plate was washed and the Fab fragment of monoclonal antibody R1 (to the α subunit) attached to alkaline phosphatase was added at 1 µg/ml for 1 h at room temperature with agitation. Detection was by the amplified alkaline phosphatase assay described in *Protocol 14*. In contrast to the assay shown in *Figure 2*, activin only showed a 0.05% cross-reaction. The detection limit was 1–2 pg/ml. (Reprinted from (42) by kind permission of the publishers, Elsevier.)

antibodies. Furthermore, high-quality streptavidin-enzyme conjugates are available from several commercial sources. In this section methods of directly labelling antibodies with enzymes will be considered. One advantage of using such antibodies is the possibility of reducing the number of stages in the subsequent ELISA.

Procedures to couple peroxidase to antibodies using glutaraldehyde or periodate have been described in detail by others (26), as have methods of coupling antibodies to alkaline phosphatase using glutaraldehyde or to β-galactosidase using maleimidobenzoyl-*N*-hydroxysuccinamide ester (26). The use of such conjugates in *Protocols 12, 13, 14,* or *15* could allow a sensitive two-site ELISA to be conducted in which an enzyme-labelled antibody re-placed the corresponding biotinylated antibody, the subsequent incubation with streptavidin-enzyme being then omitted.

Although the above procedures may give satisfactory assays one problem commonly encountered with home-made conjugates is that they may contain a heterogeneous mixture of molecular species and exhibit enhanced non-specific binding. This may mean there may be no value in applying fluoro-metric or amplified enzyme assays since no sensitivity gain can be achieved. The best method of coupling antibodies to enzymes to produce defined conjugates exhibiting low non-specific binding is to couple Fab antibody fragments through their thiol group to enzymes activated by a heterobifunc-tional reagent followed by size fractionation to exclude aggregated material.

Ishikawa and co-workers have described in detail the preparation of such conjugates from polyclonal antibodies (40), and their use in ultrasensitive ELISAs. The procedures can readily be adapted for Fab fragments prepared from mouse IgG1 and IgG2a monoclonal antibodies via pepsin digestion followed by mild reduction, although IgG2b monoclonals are too sensitive to pepsin digestion to allow this approach to be used. An alternative, although more technically demanding, route to Fab fragments of any mouse monoclonal antibody is to clone cDNA from the cells and express the Fab or $F(ab)_2$ fragments in bacteria.

An additional procedure to simplify two-site assays is to carry them out as one-step rather than two-step procedures. For example, the enzyme-labelled antibody can be added to the antibody-coated microplate well just before the sample containing the analyte. After a period of incubation the plate is washed and substrate added. One disadvantage of this approach is that samples containing very large amounts of analyte may give less signal than those with smaller amounts. This is known as the high-dose 'hook effect'. Another disadvantage is that substances in the sample may inactivate the enzyme used as label. Despite these potential difficulties many commercially available kits use one-step assay procedures. For research applications two- or three-step assays are more commonly used and are less likely to give problems.

3.4.5 Stable immobilization of antibody on microplates

Passive adsorption of antibodies to plastic is by far the most widely used method of preparing microplates for use in two-site ELISAs. However, some antibodies lose much of their functional binding affinity and capacity when adsorbed to plastic and the detachment of the antibody during the assay can impair sensitivity and precision. An alternative procedure makes use of the carbohydrate residues found in the Fc of antibodies (43). The antibody is reacted with sodium periodate at pH 5 to generate aldehyde groups. After removal of excess reagents the diluted antibody is added to a 96-well microplate carrying hydrazide groups on its surface, for example AvidPlate-HZ. *Protocol 16* shows the procedure in detail.

Protocol 16. Orientated coupling of antibodies to microplates via carbohydrate on the Fc

Reagents
- AvidPlate-HZ (BioProbe International)
- Sodium periodate (BDH 30200)
- Ethylene glycol (Sigma E9129)
- Sephadex G25 (Pharmacia)

Methods
1. Dialyse the antibody against 50 mM acetate buffer, pH 5.0. Adjust the antibody concentration to 1–2 mg/ml. (See *Protocol 11*.)

Protocol 16. *Continued*

2. Add 100 µl of 100 mM periodate solution to each 900 µl of antibody.

3. Incubate the solution in the dark for 30 min at room temperature.

4. Stop the reaction by adding 1/100 volume of ethylene glycol.

5. Desalt the antibody solution on a suitably sized column of Sephadex G25 run in 50 mM acetate buffer, pH 5.0

6. Dilute the antibody in 50 mM acetate buffer, pH 5.0, to give concentrations in the range 1–10 µg/ml.

7. Add 100 µl to each well on AvidPlate HZ.

8. Cover the plates and incubate overnight at room temperature.

9. Empty the contents of the plate and add 150 µl of a fresh solution of 0.1 M Tris–HCl buffer, pH 7.5, 1% (w/v) BSA. The solution should be filtered through a 0.45 µm filter just before use. If not needed immediately, plates can be stored in this solution with 0.1% (w/v) NaN$_3$.

10. After 1 h at room temperature wash the plate 10 times with wash solution (TBS for alkaline phosphatase and PBS for peroxidase). The plate is left soaking in the solution for an hour to remove weakly bound antibody, and is then ready for ELISA (*Protocol 12*).

Another method of coupling antibodies tightly to microplates is to make use of the biotin–streptavidin interaction. Microplates to which streptavidin is covalently attached are now available commercially (Bio-Products and Labsystems). Coating of these plates with biotinylated antibody prepared as in *Protocol 11* is straightforward. To all intents and purposes the strong interaction of biotin with streptavidin is as stable as a covalent linkage. The author's laboratory uses 50 µl/well of 2.5 µg/ml biotinylated antibody overnight in PBS, 1% (w/v) BSA, 0.1% NaN$_3$ and stores the unused plates in this solution in the cold room for several months. These plates obviously cannot be used with biotinylated detector antibodies and are best used with antibody or antibody fragments directly attached to enzymes. *Figure 4* shows a two-site assay for recombinant inhibin recently developed in our laboratory. This uses the same pair of antibodies as the β-galactosidase-based detection system employed in *Figure 2*. The new assay uses a monoclonal antibody to the inhibin βA subunit immobilized on a streptavidin-coated plate with detection using the Fab fragment of a monoclonal antibody to the α subunit coupled to alkaline phosphatase. Detection of the enzyme uses an amplified alkaline phosphatase assay (*Protocol 14*).

The author strongly recommends the use of attachment of antibodies by either of the two methods described above, as both have recently given excellent results in our assays. For many years it was not possible to demonstrate that commercially available microplates, claiming covalent attachment, delivered, in practice, the anticipated improvements in assay performance to justify their use. Problems were noted in reproducibility and the reality of covalent attachment versus adsorption was open to question. Orientated coupling through the carbohydrate in the Fc, and immobilization via the biotin–streptavidin interaction, by contrast, give a reproducible and stable immobilization offering a higher density of functional antibody than plates prepared by simple adsorption. One can confidently use them to assay serum samples without worrying that the coating antibody will be partially displaced from the plastic as is known to be the case with plates coated by passive adsorption of antibody. The inhibin ELISA described above (6), using antibodies immobilized on hydrazide microplates, was recently validated for measurement of the dimeric form of human inhibin throughout the menstrual cycle (47).

4. Immunoblotting

4.1 Electroblotting

Additional evidence that an antibody raised to a synthetic peptide immunogen actually reacts with the molecule of interest can be obtained by immunoblotting. Following SDS-PAGE and blotting most anti-peptide antibodies will react with the denatured polypeptide chains of the parent protein. The procedures for immunoblotting itself are no different when anti-peptide antibodies rather than conventionally made antibodies are used. *Protocol 17* worked well for the author on the first attempt, and has been used since.

Protocol 17. Immunoblotting using alkaline phosphatase as the enzyme label and biotin/streptavidin detection

Reagents

- Biotinylated molecular weight standards for SDS-PAGE (Sigma MWSDS100B)
- Apparatus for SDS-PAGE and electroblotting
- Nitroblue tetrazolium (Sigma N6876). Make up a stock solution containing 0.5 g in 10 ml of 70% DMF. Keep at 4°C
- Bromochloroindolyl phosphate (Sigma B6149). Make up a stock solution containing 0.5 g in 10 ml of DMF

- Nitrocellulose (Biorad blotting grade)
- Streptavidin–alkaline phosphatase (e.g. Serotec)
- Biotinylated anti-rabbit or anti-sheep IgG (e.g. Serotec)
- Tris-buffered saline (TBS) 50 mM Tris–HCl buffer, pH 7.5, 0.15 M NaCl
- Alkaline phosphatase buffer: 0.1 M Tris–HCl, pH 9.5, 0.1 M NaCl and 5 mM $MgCl_2$

Method

1. Carry out electrophoresis on tissue or cell samples thought to contain the protein of interest. Samples can be heated to 100°C with or without the inclusion of 2-mercaptoethanol. Load molecular weight standards in adjacent lanes.[a]

2. Blot the gel on to nitrocellulose using whatever electroblotting apparatus is available.

Protocol 17. *Continued*

3. Rinse the nitrocellulose thoroughly with TBS (two changes of 5 min each).

4. Block excess sites on the nitrocellulose by gently shaking it in TBS, 1% (w/v) BSA for 1 h at room temperature.

5. React the nitrocellulose with anti-peptide antibody:

    (a) monoclonal antibodies or affinity-purified polyclonal antibodies: incubate the nitrocellulose with 1–10 μg/ml of biotinylated antibody in TBS, 1% (w/v) BSA for 1 h at room temperature. Shake the solution gently during this time;

    (b) polyclonal antiserum: incubate the nitrocellulose with a 10^{-2}–10^{-4} dilution of serum in TBS, 1% (w/v) BSA for 1 h at room temperature

6. Wash the nitrocellulose thoroughly with TBS (four changes of 5 min each).

7. Alternative procedures are needed depending on the nature of the primary antibody:

    (a) If biotinylated monoclonal antibody or biotinylated polyclonal antibodies were used in step **5**: incubate the nitrocellulose with a 10^{-3} dilution of streptavidin–alkaline phosphatase made up in TBS, 1% (w/v) BSA. A 1 h incubation with agitation should be sufficient.

    (b) If diluted crude antisera were used in step **5**: add a biotinylated second antibody. Dilute the biotinylated affinity-purified anti-rabbit or anti-sheep IgG to 10^{-3} in TBS, 1% (w/v) BSA and incubate with the nitrocellulose for 1 h at room temperature with agitation. Wash the nitrocellulose with TBS (four changes of 5 min each). Then incubate with a 10^{-3} dilution of streptavidin–alkaline phosphatase as above.

8. Wash the nitrocellulose with TBS (six changes of 5 min each).

9. While the nitrocellulose is washing make up the substrate solution as follows: to 10 ml of alkaline phosphatase buffer add 50 μl of nitroblue tetrazolium stock and 50 μl of bromochloroindolyl phosphate stock.

10. Incubate the nitrocellulose in this solution. Purple bands will form quite quickly; gently agitate the reaction vessel until background staining begins to rise. Be sure to turn the nitrocellulose over in case your bands are on the other side!

11. Stop the reaction by rinsing thoroughly with water. The nitrocellulose strips can be stored dry and photograph well.

[a] The use of biotinylated standards has an advantage that unknowns and molecular weight standards are adjacent on the stained nitrocellulose and molecular weight calculations for unknowns are easier to make.

Figure 5 shows examples of immunoblots prepared in our laboratory using this protocol to detect anti-peptide monoclonal antibodies to inhibin.

5. Immunocytochemistry

Techniques for immunocytochemistry using anti-peptide antibodies are no different from those using conventionally made antibodies and this is not the place to discuss them in detail. There is no doubt that anti-peptide antibodies have proved extremely useful for immunocytochemistry for several reasons:

(a) They allow antibodies to be made to molecules within weeks of gene sequences being available.

(b) They often react with conserved epitopes which allow the same antibodies to be used to stain homologous structures across a wide evolutionary spectrum (44).

(c) They stain fixed and wax-embedded tissues more commonly than conventionally made antibodies whose epitopes are more conformational in nature.

These points can illustrated by monoclonal antibody (E4) to the βA subunit, found in inhibins and activins, which has been referred to in the sections on two-site assays (*Figures 2, 3* and *4*) and immunoblotting (*Figure 5*). This antibody has proved very useful for the detection of the βA subunit in a wide range of tissues including ovary, placenta, testis, and in early embryos of mouse and zebra fish where, as a component of activins, it is thought to be involved in inductive interactions (45). This monoclonal antibody was raised to a synthetic peptide corresponding to sequence 82–114 of the human βA subunit. Recently we have used shorter peptides within this region to narrow down the epitope to the region 89–97 with the AspAsp sequence at 95–96 being particularly important. This is deduced from the weak reactivity of the antibody with the βA subunit from zebra fish which has substitutions at these sites. That the sequence of the epitope SMLYDDG contains no lysine or histidine may account for the resistance of the epitope to paraformaldehyde fixation.

Figure 6 shows staining of mouse ovarian tissue with the E4 antibody and the legend describes how the specimen was processed. Detailed procedures for immunocytochemical staining are described elsewhere (46). Of particular importance are methods used to confirm the specificity of the staining patterns observed (46).

6. Summary

(a) Antibodies to peptides are extremely useful reagents, particularly when conventionally made antibodies to a molecule of interest are not available.

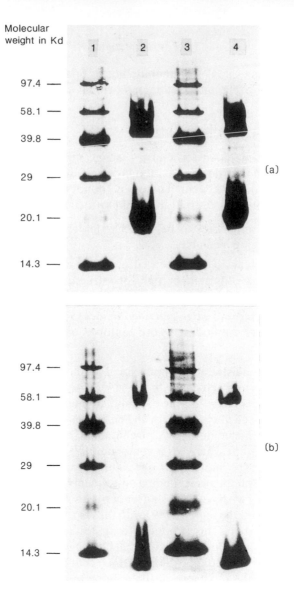

Figure 5. Immunoblotting of inhibin molecular forms with monoclonal antibodies raised to synthetic peptides. Inhibin molecular forms were concentrated by affinity chromatography using a monoclonal antibody to the α subunit (R1). The eluted samples were heated with SDS and 2-mercaptoethanol and analysed by SDS-PAGE and immunoblotting. (a) Shows that the biotinylated monoclonal antibody (R1) to the inhibin α subunit stains bands of apparent molecular weight 20 kDa, 44 kDa and 55 kDa in bovine inhibin concentrates (lanes 2 and 4) by reference to the biotinylated standards in lanes 1 and 3. (b) Shows that the biotinylated monoclonal antibody (E4) to the inhibin βA subunit stains bands of apparent molecular weight at 13 kDa and 58 kDa in the same inhibin concentrates as in *Figure 5a* (lanes 2 and 4). The monoclonal antibody R1 was prepared by immunization of mice with a synthetic peptide corresponding to the 1–32 sequence of the 20 kDa inhibin α subunit (28), and the monoclonal antibody E4 had been prepared by immunization of mice with the 82–114 sequence of the βA subunit of 32 kDa human inhibin.

176

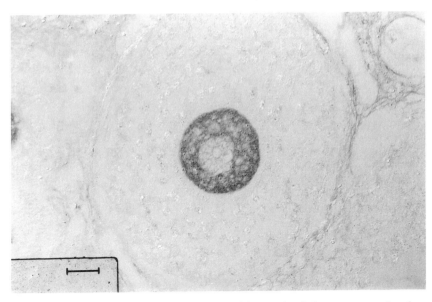

Figure 6. Immunolocalization of the inhibin/activin β subunit in mouse ovarian tissue. The figure shows an antral follicle with weak staining on granulosa cells and strong staining in the oocyte cytoplasm. Scale bar is 25 μm. Mouse ovaries were fixed in 4% (w/v) paraformaldehyde overnight, embedded in wax and cut into 5 μm sections. The sections were deparaffinized with xylene and rehydrated through an alcohol series. After removal of endogenous peroxidase by incubation with 0.1% w/v H_2O_2 in PBS for 30 min the sections were washed twice in PBS for 5 min and permeabilized with PBS, 0.5% Triton X100 for 10 min. The sections were washed again in PBS, blocked with 10% w/v normal sheep serum for 30 min, and then incubated overnight at 4°C with the E4 monoclonal antibody (to the inhibin βA subunit) at 10 μg/ml diluted in PBS, 1% sheep serum. The sections were washed three times, for 10 min each time, with PBS before staining with a Vectastain ABC kit (Vector Laboratories) according to the manufacturer's instructions. Full details have been reported elsewhere (45).

(b) Antibodies useful for blotting the protein of interest can almost always be obtained, with a good chance that the same antibodies will be useful for immunocytochemistry.

(c) Antibodies with high affinity for the parent protein suitable for use in immunoassay are most commonly observed following immunization with N- or C-terminal sequences.

(d) It is sometimes possible to make useful high-affinity monoclonal antibodies to internal sequences, by rigorous screening of the fusion against the parent protein, even though such antibodies are a minor proportion of the polyclonal population.

(e) As with all antibodies, unexpected cross-reactions can occur with antipeptide antibodies and specificity must be proved to be adequate for the purpose for which they are used.

Nigel P. Groome

Acknowledgements

The author is grateful to the UK Medical Research Council, the Multiple Sclerosis Society of Great Britain and Northern Ireland, and the Cancer Research Campaign for financial support of his laboratory.

References

1. Pelton, R. W., Saxena, B., Jones, M., Moses, H. L., and Gold, L. I. (1991). *J. Cell. Biol.*, **115**, 1091.
2. Kashimoto, A., Saito, N., and Ogita, K. (1991). In *Methods in enzymology* (ed. R. Wu), Vol. 200, pp. 447–54. Academic Press, London.
3. Tsitsiloni, O. E., Yialouris, P. P., Heimer, E. P., Felix, A. M., Evangelatos, G. P., Soteriadis-Vlahos, C., Stiakakis, J., Hannappel, E., and Haritos, A. A. (1988). *J. Immunol. Methods*, **113**, 175.
4. Duncan, M. E., McAleese, M. E., Booth, N. A., Melvin, W. T., and Fothergill, J. E. (1992). *J. Immunol. Methods*, **151**, 227.
5. Groome, N. P. and Lawrence, M. (1991). *Hybridoma*, **10**, 309.
6. Groome, N. P. (1991). *J. Immunol. Methods*, **145**, 65.
7. Muller-Berghaus, G., Scheefers-Borchel, U., Fuhge, P., Eberle, R., and Heimburger, N. (1985). *Scand. J. Clin. Lab. Invest.*, **45**, Suppl. 178, 145.
8. Barry, R., Lawrence, M., Thompson, A., McDonald, I., and Groome, N. P. (1990). *Neurochem. Int.*, **16**, 549.
9. Gnann, J. W. and Oldstone, M. B. A. (1990). *Current Topics Microbiol. Immunol.*, **160**, 131.
10. Nakamura, M. *et al.* (1991). *J. Neuroimmunol.*, **35**, 167.
11. Kasprzyk, P. G., Cuttitta, F., Avis, I., Nakanishi, Y., Treston, A., Wong, H., Walsh, J. H., and Mulshine, J. L. (1988). *Anal. Biochem.*, **174**, 224.
12. Hobbs, R. N. (1989). *J. Immunol. Methods*, **117**, 257.
13. Boudet, F., Theze, J., and Zouali, M. (1991). *J. Immunol. Methods*, **142**, 73.
14. Søndergard-Anderson, J., Lauritzen, E., Lind, K., and Holm, A. (1990). *J. Immunol. Methods*, **131**, 99.
15. Devey, M. E. and Steward, M. W. (1988). In *ELISA and other solid phase immunoassays* (ed. D. M. Kemeny and S. J. Challacombe), pp. 135–53. John Wiley, London
16. Spangler, B. D. (1991). *J. Immunol.*, **146**, 1591.
17. Mariani, M., Bracci, L., Presentini, R., Nucci, D., Neri, P., and Antoni, G. (1987). *Mol. Immunol.*, **24**, 297.
18. Schaaper, W. M. M., Lankhof, H., Puijk, W. C., and Meloen, R. H. (1989). *Mol. Immunol.*, **26**, 81.
19. Devey, M. E., Bleasdale, K., Lee, S., and Rath, S. (1988). *J. Immunol. Methods*, **106**, 119.
20. Aldwin, L. and Nitecki, D. E. (1987). *Anal. Biochem.*, **164**, 494.
21. Ivanov, V. S. Sovorova, Z. K., Tchikin, L. D., Kozhich, A. T., and Ivanov, V. T. (1992). *J. Immunol. Methods*, **153**, 229.
22. Geahlen, R. L., Loudon, G. M., Paige, L., and Lloyd, D. (1992). *Anal. Biochem.*, **202**, 68.

23. Marsden, H. S., Owsianka, A. M., Graham, S., McLean, G. W., Robertson, C. A., and Subak-Sharpe, J. H. (1992). *J. Immunol. Methods*, **147**, 65.

24. Ekins, R. P. (1974). *Br. Med. Bull.*, **30**, 3.

25. Ekins, R. P. (1981). In *Monoclonal antibodies and developments in immunoassay* (ed. A. Albertini and R. Ekins), pp. 3–21. Elsevier/North Holland, Amsterdam.

26. Harlow, E. and Lane, D. (1988). In *Antibodies: A laboratory manual*. Cold Spring Harbor Press, Cold Spring Harbor, NY.

27. Chersi, A., Ruocco, E., and Muratti, E. (1989). *J. Immunol. Methods*, **122**, 285.

28. Groome, N. P., Hancock, J., Betteridge, A., Lawrence, M., and Craven, R. (1990). *Hybridoma*, **9**, 31.

29. Bramley, T. A., Menzies, G. S., Baxter, G., Webb, R., and McNeilly, A. (1992). *J. Endocrinol.*, **134**, 341.

30. Bolton, A. E. and Hunter, W. M. (1986). In *Handbook of experimental immunology* (ed. D. M. Weir), pp. 26.1–26.56. Blackwell, Oxford.

31. Hudson, L. and Hay, F. C. (1989). *Practical immunology*, 3rd edn, pp. 49–53. Blackwell, Oxford.

32. Boscato, L. M., Egan, G. M., and Stuart, M. C. (1989). *J. Immunol. Methods*, **117**, 221.

33. Geysen, H. M., Rodda, S. J., Mason, T. M., Tribbick, T. M., and Schoofs, P. G. (1987). *J. Immunol. Methods*, **102**, 259.

34. Khachigian, L. M., Evin, G., Morgan, F. J., Owensby, D. A., and Chesterby, C. N. (1991). *J. Immunol. Methods*, **140**, 249.

35. Bobrow, M. N., Harris, T. D., Shaughnessy, K. J., and Litt, G. (1989). *J. Immunol Methods*, **125**, 279.

36. Brooks, J. L., Mirhabibollahi, B., and Kroll, R. G. (1991). *J. Immunol. Methods*, **140**, 79.

37. Johannsson, A., Stanley, C. J., and Self, C. H. (1985). *Clin. Chim. Acta*, **148**, 119.

38. Wood, W. G. (1991). *Scand. J. Clin. Lab. Invest.*, **51**, Suppl. 205, 105.

39. Jones, S. L., Cox, J. C., Sheperd, J. M., Rothel, J. S., Wood, P. R., and Radford, A. J. (1992). *J. Immunol. Methods*, **155**, 233.

40. Ishikawa, E., Imagawa, M., Hashida, S., Shinji, Y., Hamaguchi, Y., and Veno, T. (1983). *J. Immunoassay*, **4**, 209.

41. Johannsson, A. and Bates, D. L. (1988). In *ELISA and other solid phase immunoassays* (ed. D. M. Kemeny and S. J. Challacombe), pp. 85–104. John Wiley, London

42. Groome, N. P. and O' Brien, M. (1993). *J. Immunol. Methods*, **165**, 167.

43. Brillhart, K. L. and Ngo, T. T. (1991). *J. Immunol. Methods*, **144**, 19.

44. Jones, M., Cordell, J. L., Beyers, A. D., and Mason, D. Y. (1993). *J. Immunol.*, **150**, 5429.

45. Albano, R., Groome, N. P., and Smith, J. (1993). *Development*, **117**, 711.

46. Cuello, A. C. (ed.) (1983). *Immunocytochemistry* (IBRO Handbook series: Methods in the neurosciences). John Wiley, Chichester.

47. Groome, N. P., Illingworth, P. J., O'Brien, M., Cooke, I., Ganesan, T. S., Baird, D. T., and McNeilly, A. S. (1994). *Clin. Endocrinol.*, **40**, 717.

7

Epitope mapping using synthetic peptides

JANE WORTHINGTON and KEITH MORGAN

1. Introduction

It is of vital importance to researchers in many fields to identify the specific sequences on proteins (epitopes) that are recognized by antibodies or by T-cells. Three basic strategies for epitope identification are currently used. One approach is to use predictive algorithms to identify possible epitopes, to synthesize representative peptides, and then test them (see Chapter 2). This approach is limited by the ability of such programs to identify candidate sequences and the amino acid sequence of the protein must be known. A second approach is to make fragments of proteins by enzymatic methods or by recombinant DNA techniques (see Chapter 8) and to use these to localize an epitope to a particular region of a protein. The sequence of the whole protein need not necessarily be known at this stage and the sequence of the region containing the epitope(s) can subsequently be identified. A third strategy (epitope scanning), which again requires knowledge of the protein sequence, is to synthesize a number of small peptides (usually overlapping in sequence), each representing a small stretch of the protein, which together represent the complete amino acid sequence of the protein (or that part of it which is of interest). These small peptides are then tested by the appropriate assay system and specific epitopes identified.

In practice each of these strategies may be combined or each used at a specific stage of identifying epitopes. Factors such as cost, both in time and money, and the practicality of employing specific techniques in individual laboratories will all determine the approach to epitope identification. Similarly whether one is trying to identify **any** or **all** possible epitopes of a particular protein will also determine the balance of the strategies undertaken.

Epitope scanning is certainly the most comprehensive approach to identifying linear epitopes on a protein of known primary structure. Peptides for epitope scanning can be synthesized using resin-based methods but this can be costly and time-consuming where large numbers of peptides are required. These techniques produce large (mg) quantities of peptides, some of which may be of limited interest once an epitope has been identified, and are usually more appropriate for the production of large quantities of a small number of peptides. A larger number of peptides can be synthesized on resin in quantities up to 5–10 μmol per peptide using the BT 7400 Multiple Peptide System (Biotech Instruments Ltd). The main limitation of this system is the number of peptides (up to 48) which can be synthesized at one time. A major advance in epitope scanning has been the development of the technique, described by Geysen *et al.* (1), which allows the simultaneous synthesis of hundreds of peptides on specially activated polyethylene rods or pins.

The Geysen technique of multiple peptide synthesis has to date mostly been used to identify antibody epitopes, using peptides which remain on the pins where they can be repeatedly assayed. Different supports for peptide synthesis have been investigated and peptides can now be synthesized on cellulose membranes allowing different approaches to the investigation of antibody binding.

For T-cell recognition of epitopes, however, the epitopes must be presented to T-cells by antigen-presenting cells in association with their major histocompatibility complex (MHC) molecules on the cell surface. Therefore, for the experimental identification of T-cell epitopes, peptides to be tested are required free in solution (i.e. eluted from the pins) so that they can be presented in association with MHC molecules by antigen-presenting cells. Using the same chemical synthesis procedures, but with different derivatized polyethylene pins, peptides can now be cleaved from the pins and produced in soluble form appropriate for the investigation of T-cell epitopes. Such soluble peptides can also be used to coat microtitre plates for the detection of antibody epitopes.

In this chapter we will deal mainly with the pin method of synthesizing multiple peptides for identification of antibody and T-cell epitopes by scanning, but many of the methods and problems of analysis would be the same however the peptides were produced. We also briefly describe the SPOTS method of synthesizing peptides on derivatized cellulose, the Mimotope strategy using peptides synthesized on rods for determining the shape and charge of an epitope from proteins of unknown sequence, and the use of peptide libraries.

2. Multiple peptide synthesis on polyethylene pins

A strategy for epitope analysis using small-scale synthesis of peptides on specially designed polyethylene rods followed by ELISA screening for the identification of antibody epitopes was described by Geysen *et al.* (1) and is now generally known as 'Pepscan' technology. The basic principle has now been modified and developed such that it is also suitable for the detection of

T-cell epitopes (2). These techniques have since formed the basis of a range of pin technology products from Chiron Mimotopes distributed in the UK by Cambridge Research Biochemicals. This company currently has a monopoly on kits for this method of epitope scanning and sells them as three systems, two B-cell epitope scanning kits (an Epitope Scanning Kit and a Mimotope Design Kit) and one T-cell epitope scanning kit (the Cleavable Peptides Kit) which can be used for any procedure involving free peptides. For most laboratories without experience with organic chemistry the use of these kits for epitope scanning is the only realistic practical approach, therefore the information in this chapter is based on that assumption.

Each epitope scanning system includes computer software for generating synthesis schedules, blocks of derivatized pins on which the peptides are synthesized, and derivatized amino acids for use in the synthesis. Using the computer software, the identity of the peptides to be synthesized, the quantities of individual amino acids needed, and the daily schedules for the synthesis can easily be generated. All other chemicals for the synthesis have to be purchased separately. A computer, a fume hood, a microbalance, a shaking-table, and a sonicator are the other major pieces of equipment required.

The following procedures are based on our experience of using the methods described in the literature provided by the manufacturer and on changes, developments, and advice distributed to users of pin technology by Chiron Mimotopes in Victoria, Australia via their publication *Pinnacles*.

We will first deal with the identification of B-cell epitopes and then with T-cell epitopes using these systems. An overall strategy for epitope scanning is shown in *Figure 1* and the major steps in the production of the peptides in *Table 1*.

3. Antibody epitopes

3.1 General strategy

The Epitope Scanning Kit software package allows the entering and storage of protein sequences and the generation of synthesis schedules for any protein sequence entered into the program. In order to generate a synthesis schedule decisions must first be taken about the length of peptides to be made and the degree of overlap of peptides. By selecting the General Net (GNET) option in the program a schedule for the synthesis of every overlapping peptide of a specified length from a protein can be generated, i.e. sequential peptides offset by one amino acid. The program can be altered for larger offsets thus reducing the number of peptides that need to be produced to scan the complete sequence of the protein.

The software allows a choice of peptide size of 3–15 amino acids offset by 1–15 amino acids. We have found octamers to be the minimum length of peptide necessary to detect all linear antibody epitopes. The manufacturer

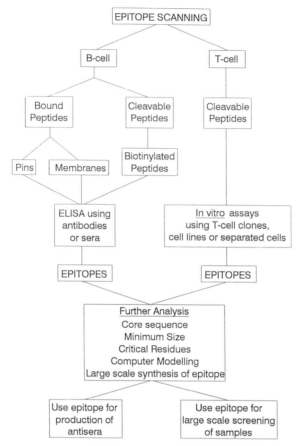

Figure 1. Epitope scanning of proteins of known primary structure. Antibody and T-cell epitopes on proteins for which the amino acid sequence is known can be identified by epitope scanning using a variety of techniques of peptide synthesis and different screening assays.

recommends octamers offset by one residue, i.e. overlapping by seven residues, as the most thorough approach to identifying linear antibody epitopes. For example:

peptide 1 QKVRRADE
peptide 2   KVRRADES
peptide 3   VRRADESC.

This would mean for a protein of 100 amino acids synthesizing 93 peptides (or 186 in duplicate), an expensive strategy when dealing with large proteins! Economies in the number of peptides to be synthesized can be made by choosing longer peptides with a smaller overlap (see *Table 2* for an example).

Table 1. Steps in the epitope scanning procedure

1. Enter protein sequence into computer and generate the synthesis schedule
2. Wash the pins and perform Fmoc deprotection
3. Prepare the amino acid solutions
4. Add Fmoc-amino acid active esters
5. Repeat steps **2–4** until peptides of desired length are synthesized
6. Acetylate terminal amino acid groups
7. Deprotect side-chains and neutralize
8. (a) For non-cleavable peptides: clean blocks and store
   (b) For cleavable peptides: cleave into appropriate solution, aliquot, and store
9. Apply suitable assay procedure and identify epitope
10. Confirm identity of epitope and investigate further

Table 2. Effect of change in peptide size and overlap on the number of peptides required for epitope scanning. This table shows an example of scanning a 100-amino-acid protein. The number of peptides needs to be doubled for duplicate copies of each peptide

Size of peptide	Offset	Number of peptides
8	1	93
9	2	47
10	3	31
11	4	24

The cost-saving benefit of choosing to synthesize longer peptides with reduced overlap is tempered by a possible loss of resolution as it may not be apparent which part of a longer peptide contains a binding sequence. Further, longer peptides may assume a secondary structure which may affect the recognition of linear epitopes. Each coupling of amino acids is less than 100% efficient, therefore, as the peptide length is increased, the proportion of final product which has the correct sequence decreases.

The synthesis schedule includes control peptides as a means of monitoring the synthesis. However, the recommended control sequences are only four amino acids long, therefore, we recommend synthesizing a number of control peptides beginning on different days such that each day of synthesis is monitored in a control peptide. An antibody which is positive for one of the control peptides but negative for the other is included in the B-cell kit. We also recommend synthesizing duplicate copies of peptides, as in our experience this is essential to ensure confidence in interpreting ELISA (enzyme-linked immunosorbent assay) results.

Having entered the protein sequence and chosen the size of peptides, overlap, etc. the program can be run to produce a 'synthesis schedule'. For each cycle of the synthesis, the weight of each amino acid required, the volume in which each should be dissolved, and a listing of which wells each amino acid should be pipetted into is produced. The pipetting information is not in an easy-to-use format and production of a simple colour-coded plan is a useful aid to accurate pipetting. Alternatively Labsystems has produced the Epiguide, a machine to guide the dispensing of the amino acid solutions into the reaction trays. A computer-generated file controls a light emitting diode under each well of the reaction tray by lighting up to indicate the wells that should have the particular amino acid being dispensed. It can handle up to 10 reaction trays and works as a stand-alone instrument after the transfer of a command file from a computer. The multipin synthesis software creates files suitable for the Epiguide but these can also be created by the user allowing its use with other microplate assays. The cost of this instrument in February 1992 was about US$7000 which may be prohibitive for most laboratories.

3.1.1 The pins

Peptides are synthesized on polyethylene pins which are inert but have a detachable crown which has been grafted with acrylic acid by radiation. Mono-*t*-butyl oxycarbonyl-1,6-diaminohexane is then coupled to the grafted polyacrylic acid matrix and after removal of the temporary *t*-butyloxycarbonyl amino protecting group, β-alanine is added as a spacer group. All peptides will thus have an alanine residue at the carboxy terminus (see *Figure 2*). The crown forms a defined active surface on which peptides can be synthesized and is designed to have a large surface area in order to maximize yield of peptide.

The pins are held in polypropylene blocks arranged in the same format as 96-well microtitre plates. Individual pins can be removed so that unused pins are not wasted and may be added in at later stages in the synthesis should shorter peptides be required. The control pins should be stored at −20 °C and unused blocks of pins should be stored dry and as cool as practicable. For example they may be stored in a refrigerator at 4–8 °C in a sealed container with silica gel. The drying agent should not come into contact with the derivatized surface of the pins.

3.1.2 The amino acids

All amino acids used in the synthesis have their α-amino groups protected with the 9-fluorenylmethyloxycarbonyl (Fmoc) group and are in the activated form as their pentafluorophenyl esters (with the exception of serine and threonine which are supplied as oxo-benzotriazine esters). Side-chain groups are also protected; *t*-butyl ether for serine, threonine, and tyrosine; *t*-butyl ester for aspartic acid and glutamic acid; *t*-butyloxycarbonyl for lysine and histidine; 4-methoxy-2,3,6-trimethylbenzenesulphonyl for arginine; and trityl for cysteine. The amino acid set used for epitope scanning consists of the 20 naturally occurring L optical isomers; however, the D isomers are provided for use in certain Mimotope strategies (see Section 5.2).

(a) **Pin bound peptide**

(b) **Cleavable peptide**

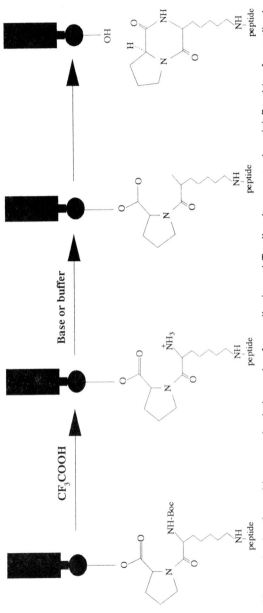

Figure 2. The synthesis of peptides on polyethylene pins for antibody and T-cell epitope scanning. (a) Peptides for antibody epitope scanning are synthesized on to polyethylene pins, via a linker (HMD, hexamethyldiamine) as described in Section 3.1.3, where they remain and can be repeatedly scanned with antibodies. (b) Peptides for T-cell epitope scanning are synthesized on to polyethylene pins via a cleavable linker (Section 8.2). On completion of synthesis peptides are cleaved under mild base conditions.

Derivatized amino acids should be stored cold and dry, for example in a sealed container within a sealable bag containing silica gel in a $-20\,^{\circ}$C freezer.

3.1.3 Chemistry of synthesis

Peptides are synthesized, from the carboxy terminus, on the pins by a repetitive cycle of Fmoc deprotection initially of the immobilized β-alanine, washing, and coupling of the next amino acid in its protected and activated form, to the newly exposed amino group. Synthesis usually proceeds at a rate of one amino acid per day; however, coupling time can be reduced from overnight to 4 h without detrimental effect, allowing a faster rate of synthesis. After removal of the final Fmoc group, the terminal amino acid is capped by acetylation and side-chain protecting groups are removed by trifluoroacetic acid containing scavengers (see *Figure 2*). Pins are thoroughly washed before use to remove contaminants.

The chemistry of peptide synthesis has been simplified in order to achieve simultaneous multiple synthesis and consequently certain reactions may proceed at less than optimal conditions. Under such conditions some peptides will be less efficiently synthesized and may be under-represented in subsequent assays. The manufacturer estimates that 80% of peptides are greater than 80% pure. It should also be noted that as the peptides are synthesized from the carboxy terminus the C-terminal residues may not be free for antibody binding.

It is difficult to estimate the cost of epitope scanning additional to that of the original kit, as the use of solvents and other solutions will depend on the length of peptides synthesized, the size of containers used at washing stages, the number of blocks used in any one synthesis, and a number of other factors. We have found the cost to be as much as 30–50% of the original kit if all the blocks are used in three or four different syntheses.

Protocol 1. Multiple peptide synthesis on polyethylene pins

Materials

- Epitope Scanning Kit containing: Fmoc-amino acid active esters, blocks of pins including control pins, reaction trays, 1-hydroxybenzotriazole (HOBt) (Cambridge Research Biochemicals)
- *N,N*-dimethylformamide (DMF) (Peptide and AR Grade)[a]
- Methanol
- Glass scintillation vials
- Polypropylene sandwich boxes
- Computer
- Microbalance
- Fume hood
- Shaking-table
- Water baths

Method

1. Generate the synthesis schedule. Prepare an appropriate number of blocks, removing pins which are not required or upon which synthesis of shorter peptides will begin later in the schedule. Always wear clean surgical gloves when handling pins and blocks.

---

2. Use a microbalance to weigh out amino acid derivatives, according to the synthesis schedule, into clean glass sealable containers (e.g. glass scintillation vials). Derivatized amino acids should be stored at $-20\,°C$ in a sealed container in the presence of silica gel and should be allowed to stand at room temperature for about 30 min before opening to prevent hydration.

3. Prepare the HOBt solution according to the schedule.[b] HOBt readily dissolves in DMF with gentle agitation. Use glass pipettes with DMF solutions.

4. Deprotect the pins by following *Protocol 2* up to the end of step **5**.

5. Whilst the blocks are air drying at step **5** of *Protocol 2*, prepare the amino acid solutions by adding the appropriate volume of HOBt solution (recommended concentrations: 30 mM for non-cleavable peptides and 150 mM for cleavable peptides).[c]

6. Using either the print-out of the synthesis schedule or a template of each block as a guide, dispense 150 μl of amino acid solutions into the appropriate wells of the reaction trays. Using a template and dispensing one amino acid solution at a time we have found it easy to avoid mistakes.

7. Complete the deprotection procedure (*Protocol 2*, step **6**).

8. Position blocks into wells of reaction trays in the correct orientation. Blocks should be uniquely marked by scratching a label on to the surface of the block. Incubate blocks overnight at $30\,°C$ in a sealed container or plastic bag to prevent evaporation. Serine and threonine solutions will become yellow upon coupling, this is quite normal.

9. Place blocks in DMF bath and wash with agitation for 2 min.[d]

10. Remove blocks, shake off excess liquid, and wash in methanol bath for 2 min.

11. Repeat methanol wash twice more using fresh methanol. Allow blocks to air dry for 10 min. (It is possible to store blocks at this point.)

12. Repeat steps **2–11** until all peptides are assembled (or **4–11** if amino acid solutions are ready as frozen aliquots).

13. Acetylate N-termini of peptides (see *Protocol 3*).

14. Deprotect the side chains (see *Protocol 4*).

15. Blocks should be cleaned by sonication and disruption (see *Protocol 7*) before embarking upon the first antibody assay.

[a] DMF used in peptide synthesis must be free from amines and all DMF should be tested before use (method in manufacturer's protocol book). AR DMF should be treated with activated 4A molecular sieves (20 g/litre, left to stand for 48 h then decanted from sieves and retreated a second time). Peptide-grade DMF should be used in the pre-coupling soak and the coupling mixture; treated AR DMF is sufficient for other stages.

Protocol 1. *Continued*

[b] All procedures should be carried out in a fume cupboard. The large volumes of organic solvents needed for this technique should be used and disposed of according to local regulations.

[c] The amino acid solutions should not stand at room temperature for too long therefore we dissolve three amino acids at a time, dispense these, and then dissolve three more. Arginine should be dispensed last of all and proline takes the longest time to dissolve. Alternatively the amino acid solutions as required for a complete synthesis can be prepared and stored as 150 μl aliquots at −70 °C for up to 2 months with no significant loss of reactivity.

[d] We have found that a series of polypropylene sandwich boxes, each big enough to hold two blocks, are suitable for the washing stages. **Do not use PVC boxes**. Use different sets of boxes for different solutions so that all the blocks can be transferred quickly at each stage. The boxes can be placed on a shaking-table to improve the efficiency of the washing stages and fresh solvent should be used for each washing stage.

Protocol 2. Fmoc deprotection and washing [a]

Materials

- 20% piperidine in DMF
- Methanol

Method

1. Place the blocks in a bath containing 20% (v/v) piperidine in DMF, up to half the height of the pins, for 30 min. [b]

2. Remove the blocks, shake off excess liquid, and wash the pins in DMF to half the pin height for 5 min.

3. Remove blocks, shake off excess liquid, and immerse blocks **completely** in methanol for 2 min to ensure removal of piperidine.

4. Place blocks in a second methanol bath for 5 min up to half the pin height. Repeat this step in a fresh methanol bath.

5. Remove the blocks and allow them to air dry for a minimum of 10 min (e.g. while the amino acids are being pipetted). [c]

6. Wash the blocks in fresh DMF before positioning in the amino acid solutions.

[a] Pins should only be deprotected immediately before coupling of amino acids, so if peptides of varying lengths are to be prepared in the same schedule, add the pins to the blocks at the appropriate time.

[b] Carry out all procedures at room temperature.

[c] It is important in step **5** that the methanol **completely** evaporates as residual methanol will interfere with the synthesis procedure.

Protocol 3. Acetylation of N-termini of peptides

Materials

- DMF
- Acetic anhydride
- Triethylamine
- Methanol

Method

1. After addition of the final amino acids, deprotect and wash blocks (*Protocol 2*).

2. Place blocks in wells of reaction trays containing 150 μl of the acetylation mixture; DMF:acetic anhydride:triethylamine 5:2:1 (by volume).

3. Incubate for 90 min at 30 °C, in sealed plastic bags.

4. Wash blocks in a DMF bath (1 × 2 min) and methanol (3 × 2 min).

5. Air dry blocks for at least 10 min.

Protocol 4. Side-chain deprotection and neutralization [a]

Materials

- Trifluoroacetic acid (TFA) [a]
- Ethanedithiol (EDT)
- Anisole
- Methanol

Method

1. Incubate the blocks to half the pin height in a mixture of TFA:anisole: EDT 95:2.5:2.5 (by volume) for at least 4 h at room temperature (if there are arginine residues this incubation should be overnight). TFA:anisole (95:5, v/v) or TFA:water (95:5, v/v) can be used if the peptides contain no cysteine residues, with the TFA:anisole mixture being the recommended one.

2. Air dry blocks in the fume hood for 10 min.

3. Sonicate blocks in 0.1% HCl in methanol/distilled water (1:1, v/v) for 15 min.

4. Air dry blocks and store in airtight containers in the presence of silica gel.

[a] **Caution**. It is very important that this procedure is carried out in a fume hood as the solutions are corrosive, toxic, and very smelly.
[b] **Caution**. The TFA and the TFA-containing mixtures must **never** be mixed with the other wash solutions nor with the deprotection or coupling solutions as TFA and DMF react in a highly exothermic way.

3.1.4 Assessment of efficiency of the synthesis

The success of the synthesis of peptides can be monitored by comparing binding of control antisera to the presynthesized positive and negative control pins with binding to control peptides prepared in your own synthesis. Problems such as significantly different binding to presynthesized controls and your controls should be investigated. Peptides from the pins can be analysed qualitatively and quantitatively by conventional means (a service offered by Cambridge Research Biochemicals) and thus problems may be pinpointed to a particular cycle or to a consistently low efficiency of coupling.

A major problem with the technique of epitope scanning is that each individual peptide is not checked, so whilst the overall efficiency of couplings is monitored in control peptides, problems such as pipetting errors, i.e. coupling of the wrong amino acid, or very low coupling of particularly difficult residues, for example arginine and tryptophan, is not assessed. This is generally accepted as a limitation of the technique; however, it does emphasize the importance of using other techniques for further analysis of identified epitopes. This may take the form of chemical analysis of peptide from reactive pins, or synthesis of the peptide of interest by conventional means and retesting of antisera in a conventional ELISA.

3.1.5 Identification of antibody epitopes by ELISA

The assessment of antibody binding to the peptides (for the assessment of synthesis and for the detection of antibody epitopes) is carried out by a modified ELISA technique. Pins are incubated in wells of a series of microtitre plates containing firstly a blocking agent, secondly the primary antibody, thirdly an enzyme-conjugated second antibody, and finally the enzyme–substrate solution in which colour development will be proportional to the amount of primary antibody bound. Results of the assay can be determined by measuring absorbance on a microtitre plate reader and transferred directly to computer for analysis. This ELISA has additional costs compared with standard protocols as a series of microtitre plates are used for each block of pins and the pins have to be cleaned by a disruption procedure after each assay.

i. Controls

The success of the ELISA can be measured by the level of binding of control antibody to the positive control peptide and the low binding of control antibody to the negative control. Inclusion of control peptides in each ELISA affords a means of assessing interplate and interassay variation and also allows a measure of the slow but inevitable loss of peptide from the pins as a result of repeated use. The manufacturer estimates the life of the pins to be 50 assays; however, this does vary and each set of pins should be judged independently.

Before performing an ELISA with the test antisera the second antibody should first be tested in the absence of primary antibody for non-specific binding to peptides. Enzyme-conjugated antibodies which bind to the peptides are not suitable; however, we have never found this to be a problem. When testing immune sera for peptide binding it is important to first test the binding of a sample of pre-immune sera and if possible also test an irrelevant immune serum, i.e. one raised against a different antigen. Providing these controls do not bind to the test peptides, any positive binding of test immune sera should represent specific recognition of an epitope. The concentration of primary antibody used in an epitope-scanning ELISA will depend on the individual preparation. As a general rule antibodies should be used at the same concentration as used in an ELISA for the parent protein. We have used monoclonal antibodies at 1/1000–1/10 000, animal antisera at 1/500–1/1000 and human autoimmune sera at 1/1000.

Protocol 5. ELISA of pin-bound peptides

Materials

- 96-well microtitre plates suitable for ELISA (one for each stage of the ELISA, i.e. four for each block of pins)
- Microtitre plate reader
- Blocks of pins with synthesized peptides
- Test antibody
- Enzyme-conjugated second antibody (horseradish peroxidase is recommended)
- PBS (20 mM sodium phosphate buffer, pH 7.2, 0.15 M NaCl) 0.05% Tween 20 (PBST)
- Substrate solution freshly prepared 0.5 g/litre 2,2′-azinobis (3-ethylbenzthiazoline-sulphonic acid) in 0.02 M disodium hydrogen orthophosphate, 0.08 M citric acid, pH 4.0, plus 0.3 μ/ml of 30% hydrogen peroxide added immediately prior to use
- Blocking buffer (1% ovalbumin, 1% bovine serum albumin (BSA) in PBS, 0.1% Tween 20)

Method

1. Incubate pins in microtitre plates containing 200 μl/well of blocking buffer (containing 0.05% NaN$_3$) for 1 h at room temperature with gentle shaking.

2. Incubate the pins in microtitre plates containing 175 μl/well of test antisera at appropriate dilution in blocking buffer containing 0.05% NaN$_3$ (e.g. 1/1000) or diluted control antibody in the case of control pins, at 4 °C overnight. This incubation step can be carried out at 37 °C for 1 h; however, this risks damage to peptides by proteases.

3. Wash pins in PBST at room temperature with agitation (4 × 10 min).

4. Incubate pins in microtitre plates containing 175 μl/well of enzyme-conjugated second antibody at the optimal dilution (e.g. 1/2500) (see *Protocol 6*) in blocking buffer *without* NaN$_3$,[a] for 1 h at room temperature. Different second antibodies may be required for the control and test sera (see *Protocol 6*).

193

Protocol 5. *Continued*

5. Wash the pins with agitation in PBST at room temperature (4 × 10 min).

6. Incubate the pins in microtitre plates containing 150 μl/well of substrate solution in the dark at room temperature and with agitation.

7. Stop the colour development reaction by the removal of pins when sufficient colour has developed. Read the plates immediately on a microplate reader at 405 nm. The results can be read straight into the epitope scanning program ready for analysis.

8. After each assay clean the pins by sonication and disruption (see *Protocol 7*) and store in a dry and cool environment.

9. It is recommended that in order to check the efficiency of cleaning of the pins, a control experiment is performed between each assay in which the pins are incubated only with second antibody. Whilst this is an essential control experiment for any combination of pins and second antibody, we have not found it to be necessary between every assay.

^{a} NaN$_3$ can inhibit the activity of peroxidase-conjugated antibodies.

Protocol 6. Optimization of second antibody concentration

Materials

- Microtitre plates and reader
- Test antibody in 0.1 M carbonate/bicarbonate buffer, pH 9.6
- Blocking buffer (see *Protocol 5*)
- Enzyme-conjugated second antibody (see *Protocol 5*)
- PBST (see *Protocol 5*)
- Substrate solution (see *Protocol 5*)

Method

1. Coat an ELISA plate with serial dilutions (e.g. 1/50–1/10 000) of the test antibody for 1 h at room temperature.

2. Wash the plates with PBST.

3. Incubate with blocking buffer for 1 h at room temperature.

4. Incubate with serial dilutions (e.g. 1/100–1/10 000) of the enzyme-conjugated second antibody for 1 h at room temperature.

5. Wash the plates with PBST.

6. Incubate with substrate solution and after an appropriate time read the A_{405} on an ELISA reader.

7. The optimal working strength of the conjugate is determined as the dilution prior to the major drop off in signal, at a coating antibody dilution where the absorbance is on scale.

3.1.6 Cleaning of blocks of pins between assays

Peptides on pins can be assayed repeatedly (approximately 50 times), with minimal loss of binding efficiency, provided bound antibodies are completely removed prior to each use. This can be achieved by a simple sonication and disruption procedure.

Protocol 7. Cleaning of pins by disruption and sonication

Materials

- Sodium dihydrogen orthophosphate
- Sodium dodecyl sulphate (SDS)
- 2-Mercaptoethanol
- Sonic bath with an output (excluding any heater element) of at least 250 W and preferably with a frequency sweep.
- Disruption buffer: 0.1 M sodium dihydrogen orthophosphate, pH 7.2, 1% SDS, 0.1% 2-mercaptoethanol
- Methanol

Method[a]

1. Heat the disruption buffer to 60 °C (temperatures above 65 °C may be damaging to peptides).

2. Place the blocks to be cleaned tips down in the sonication bath in enough hot disruption buffer so that the pins float with their tips 1 cm above the base of the bath.

3. Sonicate for 10 min at 60 °C.

4. Rinse the blocks twice in distilled water at 60 °C for 30 sec.

5. Totally immerse the blocks in methanol at 60 °C for at least 15 sec.
   NB: Methanol is highly flammable, heat in a water bath, do not use direct heat.

6. Allow the blocks to air dry. Store the blocks in a sealed container in the presence of silica gel at 4 °C.

7. Before discarding the disruption buffer add hydrogen peroxide (30%) at 2 ml/l.

[a] Carry out all steps in a fume hood.

3.1.7 Interpretation of ELISA results

The absorbance values from the assay can be read directly into the software analysis package or can be entered manually. A print-out is produced in the form of a listing of the sequence of each peptide tested and the corresponding absorbance values. In the case of duplicates both values are printed, not the mean value. The data are also represented graphically; however, the duplicate values are treated separately and two graphs produced. The graphics within the epitope scanning package are limited and most users will prefer to transfer the data into other more familiar and versatile packages.

3.1.8 Identification of epitopes recognized by monoclonal antibodies

When analysing monoclonal antibodies the results are usually very clear with a single area of high antibody binding spanning approximately four peptides (see *Figure 3*) and this can be confirmed as the correct interpretation using a simple algorithm contained in the computer software provided with the epitope scanning system. This algorithm takes the mean of the lowest 25% of all values, and adds three times the standard deviation and treats all values above this cut-off as significant. To assist with analysis of the data the print-out also gives information such as the 10 peptides with the highest binding recorded and the mean of the lowest 25% of the results.

The identification of monoclonal antibody epitopes represents to date the most frequently used application of multipin technology. Providing an appropriate concentration of antibody is available epitopes are usually identified; however, we have failed to identify epitopes for a number of monoclonal antibodies thought to recognize conformational epitopes.

3.1.9 Identification of epitopes recognized by polyclonal antisera

For polyclonal animal antisera a slightly more complex picture is usual with a number of areas showing increased binding (see *Figure 4*). In our experience

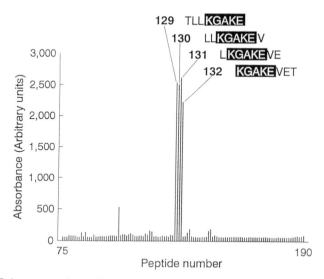

Figure 3. Epitope scanning using a monoclonal antibody. A series of peptides (8-mers, overlapping by seven amino acids) representing the hsp65 sequence of *Mycobacterium leprae* were screened with monoclonal antibodies in an ELISA. Monoclonal antibody DC16 (1/1000 dilution) recognized an epitope represented by peptides 129–132. The core sequence (KGAKE) of the epitope, common to each of the four peptides, is highlighted.

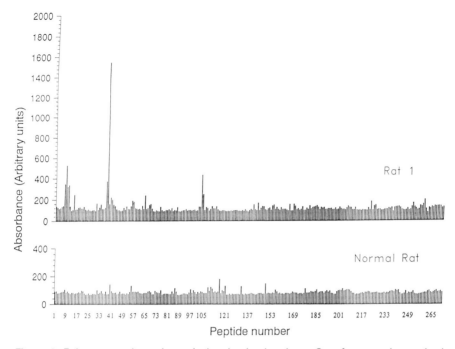

Figure 4. Epitope scanning using polyclonal animal antisera. Sera from rats immunized with type II collagen were tested for binding to peptides (8-mers, overlapping by seven amino acids) representing a portion of the type II collagen molecule. In an ELISA using sera at 1/500 dilution normal serum gave no significant binding to peptides and serum from an immunized rat detected three major epitopes.

epitopes are still relatively easy to detect and in most cases the simple algorithm described above can assist in interpretation. Using the lowest 80 or 90% of the results to determine a cut-off value may be more appropriate in some cases.

Large numbers of antibody epitopes have now been identified by epitope scanning using monoclonal antibodies or sera from immunized animals (e.g. 3–6).

3.1.10 Epitope scanning using human sera and autoimmune sera

One of the potentially most exciting applications of epitope scanning is the identification of epitopes on human autoantigens; however, there have been relatively few publications on this topic (7, 8). The reasons for this are probably two-fold; firstly, the sequences of many of the proteins of interest are as yet unknown and, secondly, the difficulties of interpreting epitope scanning data when using human sera. Human sera often exhibit high background binding to short peptides and relatively weak binding to epitopes, resulting in a decrease in the signal-to-noise ratio and increased difficulty in identifying epitopes, particularly when a number of different epitopes are recognized by the same serum sample (see *Figure 5*). In such cases increasing the serum concentration does not help and the algorithm used for monoclonal antibodies gives meaningless results.

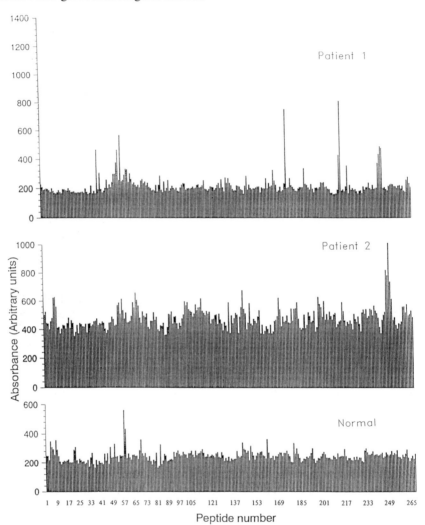

Figure 5. Epitope scanning using human autoimmune sera. Sera from rheumatoid arthritis patients with antibodies to type II collagen were tested in ELISA (1/1000 dilution) for binding to peptides representing a portion of the type II collagen molecule (the same peptides as in *Figure 4*). Background binding of normal serum was variable with significantly increased binding to one region (compare to *Figure 4*). A number of epitopes were detected by serum from patient 1. Serum from patient 2 gave very high background binding above which only one epitope can be clearly identified (peptides 249–252).

One approach to this problem is to use the data from a number of different sera and produce a consensus plot. The software for this is available from Chiron Mimotopes. This technique allows the identification of dominant or consensus epitopes but its use is not always appropriate. For example, one may be interested in identifying variations in epitope recognition by different individuals.

The type of binding profile often seen with human sera (one of high and variable background binding and no obvious peaks) may arise as a result of no major population of antibodies in the sera being directed against a linear epitope. The small peaks may result from recognition of parts of conformational epitopes or may reflect low-affinity antibody recognition of linear epitopes. The technique of epitope scanning cannot be used to determine non-linear epitopes; however, if the linear epitopes recognized by lower-affinity anti-bodies are of interest two procedures can be used to reduce background binding and enhance the result. Binding of antisera to biotinylated peptides, either in solution phase or bound to microtitre plates, appears to reduce background binding and assist in identifying epitopes (see Section 4.1).

The binding of antibody to peptide is a function of antibody concentration and antibody affinity for the peptide. High background binding may result from binding of low-affinity and relatively non-specific antibodies, present at a high concentration, and this may mask a peak of binding activity resulting from binding of a high-affinity antibody present only at low concentration. It may be possible to detect such low-concentration antibodies by performing a two-stage assay. Serum is first incubated with peptides on the pins and the bound antibodies are then eluted from the pins by a low-pH acid solution, directly into microtitre plates. The pins are then cleaned and re-incubated with the neutralized, eluted antibody solutions. Binding by the high-affinity antibodies again produces a signal; however, the low-affinity antibodies now present in a proportionally lower concentration fail to give significant binding. Thus, the signal-to-noise ratio is increased; however, the actual signal may be significantly reduced in the second assay and an amplification step may be necessary.

3.1.11 Further definition of epitopes

Depending on the information required about identified epitopes, various deductions may be made from the initial results and further investigations may be made.

i. Investigation of the core sequence

A primary scan which identifies an epitope typically shows binding to a number of overlapping peptides, for example four octamers offset by one amino acid. In the case of a monoclonal antibody the epitope can be assumed to be the linear sequence of amino acids common to each of the reactive peptides, i.e. the core sequence (see *Figure 3* and (9)). The results are more complicated from polyclonal antisera, as adjacent reactive peptides may result from different antibodies reacting with epitopes which overlap.

ii. Determination of the minimum size of an epitope

The limits of an epitope can be determined by synthesizing a series of peptides, of different lengths, representing the epitope (usually tetrapeptides to heptapeptides) and retesting the antisera thus defining the minimum length of peptides recognized by antibody (1). The Window Net (WNET) analysis option in the software can be used to produce a synthesis schedule for different sized versions of the same peptide. The minimum size of an epitope should be investigated by reducing the length of the peptide from both the carboxy and amino termini.

iii. Contribution of individual amino acid residues to antibody recognition of epitopes

For some studies it may be important to fully characterize the epitope in terms of the contribution of each amino acid to the binding of antibody (10, 11). This usually requires the synthesis of a large number of peptides and is therefore expensive. Using the Replacement Net (RNET) analysis option of the software package a synthesis schedule for peptides can be generated in which each amino acid within the peptide is replaced with some or all of the alternative residues. If antibody binding is lost when the original residue is replaced by dissimilar amino acids, this suggests that this residue contributes directly to the antibody interaction. Conversely, if antibody binding is independent of the specific nature of a residue it is thought not to interact directly with antibody but may be important as a spacer molecule (*Figure 6*). Replacement need not be restricted to the 20 genetically coded L-amino acids; unnatural D-amino acids can also be used to increase understanding about the requirements for antibody binding in terms of both residue constraints and structural requirements.

iv. Synthesis of an epitope by conventional methodology

Once an epitope has been identified, the sequence should be confirmed by measuring binding of the antibody to a conventionally synthesized peptide of that sequence (12). Differences in binding may occur as a result of conformational differences between pin-bound and plate-bound peptides. A more limited check of the epitope sequence can be performed by removing the peptide from the pin and analysing the amino acid content. This will confirm the presence of particular amino acid residues but will not provide sequence information.

v. Investigation of the biological relevance of identified epitopes

It may be of interest to assess the frequency with which an epitope is recognized. For example, one might want to screen patients' sera for binding to an

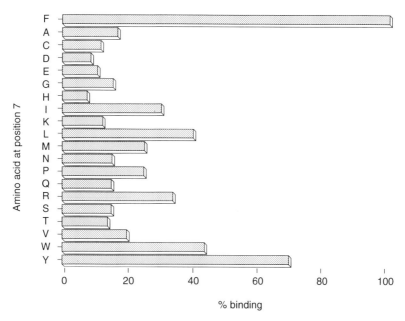

Figure 6. The effect of amino acid substitutions on antibody binding to an identified epitope. An 8-mer peptide (PAGGPGFP) from type II collagen was identified as an antibody epitope using a rat serum. Analogues of this peptide were synthesized with substitution of the phenylalanine at position 7 by the other 19 common L-amino acids. They were tested by ELISA with the same rat serum and all substitutions produced a reduction in binding compared with the parent peptide. The effect of substitutions at the other seven positions could also be studied by synthesizing 133 other analogues (i.e. 7 × 19). The *x*-axis indicates the percentage binding to each analogue compared with binding to the parent sequence.

autoantigenic epitope. In order to do this, multiple copies of peptides can be synthesized on pins and used repeatedly.

Alternatively, larger quantities of peptides could be synthesized conventionally and used to coat microtitre plates either directly or via a spacer molecule, such as biotin, to develop a solid-phase ELISA assay (see Section 4 and *Protocol 8*).

To determine the biological importance of an epitope it may be necessary to determine whether the sequence detected by epitope scanning is also recognized by antibody when in the context of native protein. This can be done in two ways: peptide can be used in a competitive assay to block antibody binding to native protein, or alternatively serum can be passed through a peptide affinity column (see Chapter 5). The unadsorbed fraction should not bind to native protein whereas the specifically eluted affinity-purified antibody should recognize the native protein. Affinity-purified antibodies could be further analysed for class (or subclass), cross-reactivity with other antigens, etc.

More information about the biological importance of an epitope may be obtained by searching protein databases for sequences in other proteins which are highly homologous with the identified epitope. Caution should be used in interpreting the relevance of such homology, especially as the smaller the sequence used, the more proteins with a similar sequence will be identified.

vi. Computer modelling

Computer programs now exist which allow the three-dimensional structure of peptides to be modelled. However, these programs require powerful computers and expertise in this field to interpret the structures. Computer modelling has been used for drug design (blocking peptides) and for modelling the interactions of peptides with receptors, antibodies, and enzymes (13).

vii. Production of anti-peptide antisera

Having identified specific epitopes, the production of polyclonal and monoclonal antibodies to the peptides may provide useful reagents for further study. Strategies for producing such antibodies are described in Chapter 4. These antibodies may be purified on peptide affinity columns (see Chapter 5).

3.2 Membrane-bound peptides and epitope scanning

A recent development in the field of solid-phase simultaneous peptide synthesis is the use of cellulose paper as a support for peptide synthesis (14). The synthesis of peptides is based on the same chemical reactions used in the pin technology and peptides attached to paper can also be reused a number of times to screen different antibodies. The potential for repetitive assays is less than the pins as the paper can only be reused about 10 times. The major advantages to this technique are in the economy in use of reagents and the speed at which peptides can be made. The volume of solvents required for washing the paper is considerably less than that required for blocks of pins. Coupling time is approximately 15 min and washing and deprotection can be completed in 30 min; thus, if so desired hexamers can be assembled in one day. This technique may be particularly useful for substitution or reduction experiments in which large numbers of peptides are required but for only a limited number of assays. This technique is available as a kit system (SPOTS: Simple Precise Original Test System) exclusively from Cambridge Research Biochemicals.

3.2.1 The membrane

Cellulose is the membrane of choice as it has an abundance of free hydroxyl groups on to which amino acids can be esterified but alternative membranes are being developed.

3.2.2 Synthesis of membrane bound peptides

An Fmoc-amino acid is esterified to the support and, after deprotection, an even distribution of amino groups is obtained. The positions on to which synthesis will take place are marked using a soft pencil (ink will be washed away by the solvents used in synthesis) and on to these, aliquots of reactive Fmoc-amino acid derivatives are dispensed. After washing, all residual amino functions on the sheet are blocked by acetylation. Removal of Fmoc-protecting groups generates free amino functions, which after staining with bromophenol blue appear as distinct spots. A dipeptide such as β-Ala–β-Ala is esterified on to the support as an anchor and spacer. This is the form in which SPOTS membranes are obtained from the supplier.

The schedule for synthesis is generated in the same way as for the synthesis of peptides on pins. Individual couplings can be monitored by the colour change of the bromophenol blue to green/yellow as the free acids react. If coupling does not appear to have taken place a second aliquot of amino acid can be added and reacted for a further 15 min. The bromophenol blue is highly sensitive and a range of colour changes is typically seen at each coupling stage. In some cases, even after a successful coupling, the blue colour will remain, particularly after the addition of valine, histidine, and isoleucine and after the coupling of proline to proline. Following the addition of the final amino acid, peptides are N-terminally acetylated and side-chains deprotected.

The normal scale of synthesis is 50 nmol, using 0.9 μl of amino acid solution per position for each cycle. However, the scale of the synthesis can be reduced by using volumes of 0.1–0.5 μl. A miniscale synthesis has now been described in which 10 nl volumes are dispensed and 100 synthesis sites are fitted on a 2 cm × 2 cm square.

3.2.3 Identification of antibody epitopes using membrane bound peptides

Membrane bound peptides are analysed by an indirect ELISA technique. BSA is a suitable blocking agent when using SPOTS, or for particularly difficult sera the blocking agent used for the pins (see *Protocol 5*) may be more appropriate. The enzyme-coupled second antibody of choice when using SPOTS is β-galactosidase-conjugated IgG, as the reaction conditions for colour development with this enzyme using the substrate 5-bromo-4-chloro-3-indoyl-β-D-galactopyranoside are very mild; thus the reusability of the paper is increased. Radioactively labelled or different enzyme-coupled second antibodies can also be used. As in other scanning procedures, the binding of second antibody to peptide should be assessed in the absence of primary antibody and the results of this borne in mind when interpreting results from a test assay. The results of the assay can either be scored by eye as positive or negative (see *Figure 7*) or can be measured densitometrically.

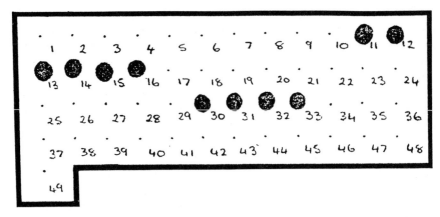

Figure 7. Epitope scanning of membrane-bound peptides with polyclonal antiserum. Peptides (10-mers overlapping by nine amino acids) representing cytomegalovirus (CMV26) were tested by ELISA using polyclonal rabbit antiserum. Two epitopes were identified: peptides 11–16 and 30–33.

All types of antibody preparation can be analysed using SPOTS but, as with all of the epitope scanning techniques, human sera tend to be the most difficult particularly because of problems of background binding. Strategies to overcome problems include trying different blocking agents, different second antibodies, and affinity purifying test antibodies.

3.2.4 Regeneration of SPOTS membranes

SPOTS membranes can be regenerated by washing the membrane in a urea, SDS, mercaptoethanol buffer, and with careful handling membranes can be probed a number of times. At the end of the regeneration procedure the membrane should be thoroughly dried with a cool hair-dryer and be stored between assays in a sealed plastic bag at −20 °C.

3.2.5 Cleavable peptides from membranes

A future development of the SPOTS technique is expected to be the elution of peptides from the membranes; however, this is not available at the time of writing. Portions of cellulose membrane containing peptide can be used directly for immunization of animals for antiserum production (15).

4. Cleavable peptides for use in identifying antibody epitopes

The production of cleavable peptides is generally associated with T-cell epitope scanning and this technology is described in detail in Section 8. However, cleavable peptides do have certain applications in the identification of antibody epitopes. Two major uses are in the analysis of 'difficult sera' such as human sera which have high background binding in classical epitope scanning assays, and as a means to analyse multiple samples for binding to a particular epitope. For these applications peptides are biotinylated prior to cleavage and can be used either in solution or solid-phase assays. Peptides can be biotinylated at the N-terminal via a spacer (typically Ser–Gly–Ser–Gly) before being cleaved from the pins. Biotinylated peptides can be made using the Cleavable Peptides Kit (Cambridge Research Biochemicals); however, unless every step is fully optimized the yield of peptides tends to be poor and investigators may prefer to take advantage of commercial services for this process.

4.1 Solution-phase assay using biotinylated peptides

One way to reduce high background binding resulting from low-affinity antibodies is to increase the affinity threshold of the assay to a point where only high-affinity antibodies are detected. This can be achieved by performing the peptide–antibody interaction in solution and, after separating bound from unbound peptide, capturing the antibody–peptide complex in a suitable manner (e.g. on to a streptavidin-coated plate) and detecting the complex via the usual enzyme-conjugated second antibody. A miniature gel filtration step is required to separate bound from unbound peptide. This is rather tedious to perform for large numbers of samples.

4.2 Solid-phase assay using biotinylated peptides

A more convenient way to achieve a comparable result to that obtained in the solution-phase assay is to first capture each of the biotinylated peptides on a streptavidin-coated plate (see *Protocol 8*). A 1/100 (or greater) dilution of peptide in cleavage buffer can be used directly to bind to streptavidin-coated plates. After washing off unbound peptide the ELISA can be completed in a conventional manner using the appropriate conjugate to detect bound antibody.

Protocol 8. Solid-phase ELISA using biotinylated peptides

Materials

- Nunc F96 Maxisorb Immunoplates (or equivalent)
- Shaking-table
- Microtitre plate reader
- Strepavidin 5 µg/ml in sterile distilled water
- Blocking solution: PBS, 2% BSA, 0.1% Tween 20[a]
- Serum buffer: PBS, 2% BSA, 0.1% Tween 20, 0.1% NaN$_3$

- Alkaline phosphatase-labelled second antibody[a]
- Substrate: *p*-nitrophenyl phosphate at 1 mg/ml in 10% diethanolamine buffer, pH 9.8
- 3 M NaOH
- PBST (*Protocol 5*)
- Cleavage buffer (*Protocol 9*)

Method

1. Coat microtitre plates with 100 µl/well of streptavidin (5 µg/µl). Leave overnight at 37 °C for the solution to evaporate to dryness.

205

Protocol 8. *Continued*

2. Wash plates three times with PBST.

3. Add 300 μl per well of blocking solution and incubate for 3 h at room temperature.

4. Wash plates three times with PBST.

5. Add 100 μl per well of appropriate biotinylated peptide solution(s) and incubate for 3 h at room temperature. (Use solution alone or irrelevant peptide for control wells.)

6. Dilute sera[b] in serum buffer, add to microtitre plates at 100 μl per well, and leave overnight at 4 °C.

7. Wash plates three times with PBST.

8. Add 100 μl per well of alkaline phosphatase-labelled secondary antibody (e.g. alkaline phosphatase anti-rat IgG at 1/1000–1/4000 dilution) in blocking buffer. Leave for 2 h at room temperature on a shaking-table.

9. Wash plates six times with distilled water.

10. Add substrate (100 μl per well) and leave for the colour to develop.

11. Stop the reaction with 100 μl per well of 3 M NaOH.

12. Read absorbance on a microplate reader at 405 nm.

[a] Alternative blocking solutions and different second antibody/enzyme–substrate combinations may be used as required.
[b] Suitable serum dilutions are typically 1/100–1/200 for human sera and 1/500–1/1000 for immune animal sera.

Both the solution-phase and solid-phase techniques can aid the detection of epitopes in difficult sera but may be more time-consuming and expensive than conventional techniques. Biotinylated peptides in a solid-phase assay may be the method of choice once an epitope of interest has been identified as this would be the fastest and cheapest way to screen large numbers of sera for reactivity to a given epitope.

5. Synthesis of peptides to mimic epitopes of proteins for which sequences are not available

Mimotopes is a strategy defined by Chiron Mimotopes for producing a peptide which mimics the original epitope recognized by an antibody. It results in the synthesis of a peptide which may not have the sequence of the epitope to which the antibody was raised (16). This strategy therefore has two advantages; it allows synthesis of epitopes from proteins for which the primary sequence is unknown but to which antibodies have been raised, and secondly, it allows the synthesis of peptides which mimic discontinuous epitopes which cannot normally be identified by epitope scanning. There are two alternative strategies for mimotope determination; however, application of these techniques has yet to be reported in the literature.

5.1 Mixture strategy based on octapeptides

In this procedure octapeptides are screened but each octapeptide is not fully defined. All possible dipeptides are synthesized ($n=400$) but they are flanked by undefined tripeptides. This is achieved by using mixtures of amino acids during coupling rather than a single amino acid solution. Thus, in theory all possible peptides including the two known residues are synthesized on each pin. Obviously the quantities of each peptide will be very small; however, the manufacturer claims that binding of antibody and peptide can be detected using a standard ELISA procedure.

Using the core dipeptide which gives the biggest signal, a second set of octamers is synthesized with one adjacent amino acid defined and the rest undefined. The triplet core giving the highest signal is then synthesized as octamers with the next amino acid defined and the rest undefined and so on until the complete epitope is defined.

5.2 Dipeptide strategy

With the second strategy all possible dipeptides are synthesized and screened. Both L and D optical isomers can be used, thus increasing the potential for variation in epitopes. The dipeptide to which antibody binds most strongly is then extended by one amino acid using the ENET procedure (see Section 6). In sequential rounds of synthesis and screening a peptide of the required length is gradually built up. Further diversity can be achieved by the use of β-alanine as a spacer residue.

6. Extension synthesis

The Extension Net (ENET) option in the computer software generates synthesis schedules for the determination of a longer reactive peptide from a known reactive peptide. The starting peptide may be a dipeptide identified in the mimotope strategy (see Section 5.2) or it may be a longer peptide. Thus ENET can be used repetitively until no further increase in activity is identified. At each round of the synthesis each of the 20 possible amino acids is added at the amino or carboxy end of the peptide. The most reactive peptide is then used as the starting point for the next round of synthesis.

7. The use of peptide libraries to identify antibody epitopes

To identify epitopes on proteins of unknown primary sequence or in cases in which the antibody recognizes a conformational epitope, an alternative to the mimotope strategy is to screen a peptide library; a collection of every possible peptide of a given length (17). The production of peptide libraries is unlikely to be an affordable approach for most laboratories, but Chiron Mimotopes have prepared and used such libraries in order to demonstrate the viability of the strategy.

Based on the finding that the average monoclonal antibody epitope is six amino acids long, their library is made up of hexamers. If one were to synthesize every possible hexamer using L-amino acids, 6.4×10^7 peptides would be required. Clearly, synthesizing this number of peptides would be a huge task and screening would almost certainly be impossible because of the amount of antiserum required. For these reasons the library is based on pools of peptides in which two adjacent amino acids are defined and the remaining four amino acids are a mixture of every possible combination. Assuming that it is reasonable to screen 2000 peptides in one assay, the peptide library is split into 2000 pools. If the first two positions are the defined amino acids each pool would contain 32 000 peptides; however, to increase the sensitivity of the system peptides are synthesized with the two defined adjacent amino acids in each of the possible five positions, thus the pools contain 160 000 peptides. Detection of the epitope within this large mixture is less of a problem than expected as not every amino acid will be critical and certain substitutions will still result in binding.

The initial assay of 2000 pools of peptides will identify the pool of peptides containing the greatest number of binders. Complete identification can be achieved either using the RNET strategy on each of the four unknown positions or by stages of synthesis and testing of mixtures of reduced degeneracy.

8. T-cell epitopes

8.1 General strategy

The stages involved in epitope scanning for T-cell epitopes are shown in *Figure 1* and *Table 1*. Similar to the B-cell epitope scanning system, the commercially available cleavable peptide system (Cambridge Research Biochemicals) has a computer program that allows the entering and storage of protein sequences and the generation of synthesis schedules. The synthesis schedules are worksheets for each cycle of the synthesis of a particular protein which is to be scanned, giving a table of weights and volumes required for the preparation of each amino acid solution. A list of every peptide to be synthesized together with the pin position on which each peptide will be found and the synthesis details are printed out.

Using the GNET synthesis option of the computer program every overlapping peptide of a specified length from a protein can be generated, i.e. sequential peptides offset by one amino acid. The program can be altered for larger offsets reducing the number of peptides that need to be produced to scan the complete sequence of a protein.

Additional programs which allow synthesis of peptides of varying length (Window Net—WNET) and the replacement of defined amino acids (ANET—Analogue Net; the equivalent of RNET in the antibody Epitope Scanning Kit) as described in Sections 3.1.11.ii and iii for antibody epitopes are equally applicable to the definition of T-cell epitopes.

The peptides for T-cell epitope scanning are synthesized on polypropylene pins by repeated cycles of Fmoc deprotection, washing, and amino acid coupling as described in Section 3.1.3. However, in contrast to antibody epitope scanning, peptides for T-cell epitope scanning must be cleaved from the pins after synthesis. To make this possible the reactive groups on the pins have been extended to incorporate a linker unit which cyclizes under alkaline conditions allowing the peptides to be cleaved from the pins (2). T-cell epitope mapping based on simultaneous multiple synthesis of different peptides has the same disadvantages described for antibody epitope scanning. As every reaction cannot be monitored and is not under optimum conditions some peptides will not be present in equivalent amounts to others. Thus there may be variable levels in background responses between peptides. Further, a particular sequence may produce a false negative result in a chosen assay method, if it is a T-cell epitope but produced suboptimally.

8.1.1 Size of peptides for T-cell epitope scanning

Most work published so far has been on epitopes recognized by CD4$^+$ T-cells with a limited amount on CD8$^+$ T-cells. The number of cells available for an assay is usually a limiting factor so initially pools of synthesized peptides can be used for testing, and then the individual peptides from a reactive pool tested to identify the epitope.

i. CD8$^+$ T-cells

Peptides of nine residues with unblocked amino- or carboxy-ends (see Section 8.2.1) are generally recommended, but the epitopes may be shorter or longer (8- to 11-mers) and thus the longer epitopes may be missed if only 9-mers are used (18). If a large sequence is being mapped the use of longer peptides to localize the region followed by testing with 8-mers, 9-mers, 10-mers, and 11-mers of that region may be a more realistic strategy.

ii. CD4$^+$ T-cells

Blocked peptides (see Section 8.2.1) of 11 amino acids or longer (12- to 15-mers) are recommended, though peptides up to 30 amino acids in length have been used. The number of peptides that need to be synthesized can be reduced by using 12-mers offset by two, or 13-mers offset by three rather than 11-mers offset by one.

8.2 Synthesis of cleavable peptides

8.2.1 The pins

The original pins used for cleavable peptides were based on the B-cell epitope polyethylene pins with the same grafted surface but with a chemical linking group enabling removal of the synthesized peptides into physiological solutions via the presence of a cleavable linker. Recently a new grafted surface has been introduced which has a co-monomer mix of methacrylic acid and dimethylacrylamide. These pins yield more peptide (about 200 nmol per pin) than the older type of pin (about 100 nmol).

After synthesis of the peptides by Fmoc chemistry the terminal amino group is capped by acetylation after the removal of the last Fmoc protecting group. The side-chain protecting groups are removed with TFA containing scavengers which remove *t*-butyl-based carbonium ions. The cleavage of peptide from the pins is initiated during the side-chain deprotection. The removal of the α-amino *t*-butyloxycarbonyl group from the lysine residue, which forms part of the linker group, enables the Lys–Pro moiety to cyclize to form a diketopiperazine and allow peptide cleavage from the pins (see *Figure 2*). All peptides will have a C-terminal extension consisting of a β-alanine spacer, and a diketopiperazine moiety attached. Thus the kit system is best for producing peptides for identifying epitopes of CD4$^+$ T-cells. Methods for producing peptides with carboxy-terminal acid or amide groups (i.e. unblocked peptides) by the multipin method using a gas-phase cleavage strategy have been developed by Chiron Mimotopes (19). These methods are not at present available in kit form but such peptides can be commercially synthesized.

The cleavable peptide blocks have blue legs fitted to differentiate them from the non-cleavable blocks which have white legs. The blocks should be stored under the same conditions as the non-cleavable blocks.

8.2.2 The amino acids

The amino acids supplied are the same derivatized amino acids as those supplied in the B-cell kit (see Section 3.1.2) but in larger quantities. They should be stored under the same conditions, i.e. dry and preferably at 4–8 °C.

8.2.3 The synthesis procedure

Peptides are synthesized on the pins from the Cleavable Peptides Kit according to *Protocol 1* steps **1–14** and *Protocols 2–4*. However, the amino acids are used at 150 mM concentration. Peptides are then cleaved by the method described in *Protocol 9*. The final yield of each peptide will be affected by peptide solubility which is dependent upon the length and sequence of the peptide. Short peptides are likely to vary more in solubility than longer peptides. Cleavage into larger volumes can help to minimize the effects of solubility differences, but if too much solution has to be added to the test cells (in order to get a high enough concentration of peptide) then the toxicity from the buffer may be a problem. Chiron Mimotopes suggest that most buffers are suitable when added at 5 μl per 200 μl of cells in tissue culture medium. They now recommend cleavage of peptides into 0.05 M Hepes buffer, pH 7.2 to 7.8 containing 20% acetonitrile (HPLC grade), but we have no experience of using this solution. The company has also described a method for the simultaneous production of peptide–carrier conjugates suitable for immunization and antibody production, using the multipin system (20).

Protocol 9. Cleavage of the peptides from the pins[a]

Materials

- Laminar flow cabinet or sterile hood
- 0.1 M citrate/phosphate buffer, pH 3.0 (sterile)
- 0.1 M sodium bicarbonate buffer, pH 8.3 (sterile)
- 0.22 μm sterile filters
- Sterile polypropylene boxes
- Sterile polypropylene reaction trays
- Sterile Eppendorf tubes

Method

1. Place the blocks of peptides in the sterile 0.1 M citrate/phosphate buffer, pH 3.0 in a closed sterile box for at least 3 h to remove residual contaminants. Alternatively a sterile reaction tray with buffer may be used for each block but the trays must also be placed inside a closed sterile box. It is important that the pH does not change as premature cleavage of the peptides will occur.

2. Cleave the peptides by placing the pins in sterile 0.1 M sodium bicarbonate buffer, pH 8.3. Dispense 150 μl of the buffer into each well of a sterile reaction tray. **Do not use plastic ELISA trays for cleavage as the peptides may stick to the plastic**.

3. Stand the pins in the solution for at least 3 h. Seal the trays inside sterile boxes. They may be left overnight to increase the yield of peptide but this procedure increases the chance of contamination.

4. **Do not filter sterilize the peptide solutions**. Store the peptide solutions frozen in the polypropylene trays with sterile lids or aliquot them into sterile Eppendorf tubes and store frozen. Repeated freezing and thawing of the peptides should be avoided.

[a] All the procedures should be carried out in a laminar flow cabinet with sterile solutions and apparatus. The 0.1 M citrate/phosphate buffer, pH 3.0 is sterilized with a 0.22 μm sterile filter. Polypropylene boxes and reaction trays are sterilized with 95% alcohol and air dried in the cabinet.

9. Screening of cleavable peptides for the identification of T-cell epitopes

The detection of T-cell epitopes is technically more demanding than identi-fication of antibody epitopes and there are a variety of options available with regard to screening the peptides. The type of assay, length of peptide used, the culture conditions, and the method of analysis for T-cell screening will be affected by a number of factors such as whether human or animal cells are being used, whether peripheral blood mononuclear cells, spleen cells, or T-cell lines or clones are being used, and whether T-helper or cytotoxic T-cells are being investigated. The availability of suitable reagents for a particular species (e.g. with cytokine release assays) or availability of detection equip-ment such as scintillation counters will also be determining factors.

Similarly the method of choice will be affected by whether the aim is to identify a single epitope or to identify all epitopes present in the protein sequence under investigation. The specific assay conditions for a particular application can only be determined by the individual investigator, but some general pointers can be given. Most published work has been on viral, bacterial, and parasite antigens (e.g. 21–23) but the technique has also been used for studying autoreactive T-cells (24).

9.1 Non-specific stimulation of T-cells and assay background

The background levels of stimulation of T-cells are dependent on a number of factors including the choice of tissue-culture medium, serum supplements, other supplements such as antibiotics, and even batch variation of supple-ments. This variation may apply to whatever read-out system is used. If one is looking at the response of a T-cell line or clone then as long as the back-ground is consistent it may not be necessary to find the conditions for a minimum background. However, reduction of background levels to a mini-mum may be particularly important when trying to identify epitopes recog-nized by peripheral blood mononuclear cells where the number of responding cells to any particular epitope may be small. In many cases the optimum conditions will have already been determined with the parent protein prior to using the peptides.

9.2 Detection systems

Most work published so far has been on epitopes recognized by CD4$^+$ T-cells with a limited amount on CD8$^+$ T-cells. At present only lines or clones can be used for mapping with CD8$^+$ T-cells. Typically peptides are incubated with 10^4 ^{51}Cr-labelled target cells in microtitre plates and with 5×10^4 effector cells in a standard 5 h ^{51}Cr release assay (25).

For CD4$^+$ T-cells, lines or clones can be used as can unexpanded lymph node and spleen cells. Successful results have also been achieved using un-enriched peripheral blood lymphocyte preparations from immune subjects, both experimental animals and patients. From published work the amounts of peptide used to produce responses have covered a wide range from 1–50 µg/ml with incubation times usually in the 3–6 day range. Most of the published work has used the uptake of tritiated thymidine as a measure of DNA synthesis and thus cell division and stimulation, whilst some workers have used the production of specific cytokines (such as IL-2) as a measure of stimulation. Many assays for cytokines in human and mouse are now available, though assays for other species are more limited.

9.3 Interpretation of results

Some workers, using tritiated thymidine uptake assays, measure stimulation indices using arbitrary cut-off points in the range of two to three times background (background levels are determined by the ^{3}H-thymidine uptake by cells with no peptide or with an irrelevant peptide). With cell lines and clones this approach seems to be satisfactory and positive responses can be clearly detected above background (*Figure 8*).

With peripheral blood mononuclear cells if there is a low frequency of responding cells, i.e. with the precursor frequency less than the number of cells in one replicate, many replicates are required to detect a statistically significant number of positive cells. More recently Chiron Mimotopes have

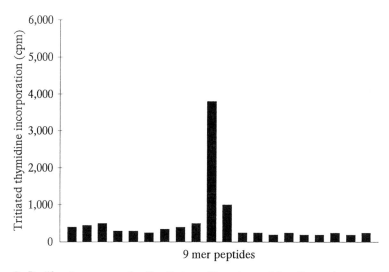

Figure 8. Proliferation assay of a T-cell clone. Twenty peptides (9-mers) representing a region of tetanus toxin containing a putative epitope were tested for their ability to stimulate a T-cell clone reactive to the whole antigen. Proliferation was measured by [^{3}H]thymidine incorporation. The epitope was clearly identified above background binding.

made a T-cell Assay Design computer program available particularly for the analysis of proliferation assays which use polyclonal T-cells such as peripheral blood mononuclear cells. The program can be used to help interpret results with T-cell clones or other assay systems which produce quantitative data.

In addition to data analysis the program can also generate a print-out to guide dispensing of reagents in the wells. A computer file of the layout is subsequently used in the analysis of the results. The analysis determines the background, cut-off, frequency of positive responses, and the statistical significance of differences between test groups. The program uses all the data of the assay and defines a normally distributed cluster of data corresponding to all the wells that have not been stimulated. An upper cut-off above which all data are scored as positive and a lower cut-off below which all data are scored as negative are determined. Values between the two cut-off values are 'indeterminant'. The frequency of positives is calculated for each test group and Poisson statistics used to determine a mean frequency for responding precursors in the cell population used. The advantages of this approach are that:

- there is no need for T-cell cloning which can be difficult, time-consuming, and which can introduce a bias in the responsive cells selected
- a wide range of specificities can be determined
- many individuals can be tested to obtain a complete picture of a set of epitopes from a protein

The disadvantages are that:

- the exact phenotype of the responding cell is not known
- human leucocyte antigen (HLA) restriction analysis is not possible
- it has not been directly compared with other methods over a wide range of antigens

This method is good for screening for epitopes (26, 27), although the reliability of the precursor frequency number has yet to be proved. An example of this type of analysis is shown in *Figure 9*.

9.4 Further definition of epitopes

As with antibody epitopes, further investigation of identified epitopes may need to be undertaken. These will include synthesizing identified epitopes by conventional methods and confirming the reactivity of the epitope and its sequence. The minimum size of the epitope and the contribution of individual amino acids may also be investigated. The WNET and ANET parts of the computer program can be used to generate synthesis schedules for the appropriate peptides.

Searching computer databases for proteins containing similar sequences may be performed and the cross-reactivity to these proteins by the T-cells under investigation can be tested.

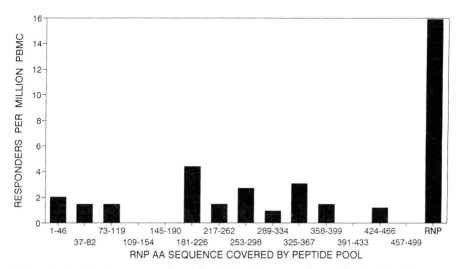

Figure 9. Proliferation assay of peripheral blood mononuclear cells. Peptides (13-mers offset by three residues) corresponding to ribonuclear protein (RNP) were pooled in groups of 11 or 12 peptides. Each pool was tested for ability to stimulate 24 replicate wells of mononuclear cells and the frequency of responding T-cells was calculated from the number of responding wells, using Poisson statistics. The RNP controls, tested with 24 replicates, were all positive, and as an exact value for the Poisson mean cannot be calculated the value shown is the minimum estimate.

10. General summary and future developments

The availability of commercial kits for antibody and T-cell scanning based on multipin solid-phase peptide synthesis, requiring only basic skills in organic chemistry, has resulted in a dramatic increase in the identification of epitopes in a wide variety of proteins by researchers from many different fields. Modifications in the type of solid support, the activated surfaces of the pins, the introduction of different linker molecules, and changes in the chemistry have improved and will continue to improve the purity and yield of peptides produced and the forms of peptide available (e.g. different chemical terminal groups and directly linked carriers and conjugates). In particular, Chiron Mimotopes have recently introduced a new pin which has five independent, grafted and derivitized supports on the same stem, which allows cleavable and non-cleavable peptides to be synthesized at the same time. Thus the range of applications for epitope scanning continues to expand.

Users of these systems should remain aware of the limitations, such as the lack of quality control of individual peptides within a synthesis, the production of false positive or negative results, and problems of analysis when using complex samples such as cells or sera from humans with autoimmune disease.

Acknowledgements

We would like to thank Dr S. J. Rodda of Chiron Mimotopes for providing the data for *Figures 8* and *9* and for helpful advice during the preparation of this chapter. Thanks also to K. M. Price of Cambridge Research Biochemicals for *Figure 7* and for permission to use information from the manufacturer's handbook.

References

1. Geysen, H. M., Rodda S. J., Mason, T. J., Tribbick, G., and Schoofs, P. G. (1987). *J. Immunol. Methods*, **102**, 259.
2. Maeji, N. J., Bray A. M., and Geysen H. M. (1990). *J. Immunol. Methods*, **134**, 23.
3. Conlan, J. W., Clarke, I. N., and Ward, M. E. (1988). *Mol. Microbiol.*, **2**, 673.
4. Hajeer, A. H., Worthington, J., Morgan, K., and Bernstein, R. M. (1992). *Clin. Exp. Immunol.*, **89**, 115.
5. Worthington, J., Brass, A., and Morgan, K. (1991). *Autoimmunity*, **10**, 201.
6. Das, M. K. and Lindstrom, J. (1991). *Biochemistry*, **30**, 2470.
7. Hay, F. C., Soltys, A. J., Tribbick, G., and Geysen, H. M. (1991). *Eur. J. Immunol.*, **21**, 1837.
8. Worthington, J., Turner, S., Brass, A., and Morgan, K. (1993). *Br. J. Rheumatol.*, **32**, 658.
9. Geysen, H. M. (1990). *Southeast Asian J. Trop. Med. Public Health*, **21**, 523.
10. Zhao, Y. P., Shi, B. J., Deng, H. K., Yu, D. T. Y., Hamachi, M., Hamachi, T., Zhao, D. F., Tang, S. R., Tsang, J. C., Park, S., Terasaki, P., and Tribbick, G. (1991). *Clin. Exp. Rheumatol.*, **9**, 235.
11. Geysen, H. M., Meloen, R. H., and Barteling, S. J. (1984). *Proc. Natl. Acad. Sci. USA*, **81**, 3998.
12. Novak, J., Sova, P., Krchnak, V., Hamsikova, E., Zavadova, H., and Roubal, J. (1991). *J. Gen. Virol.*, **72**, 1409.
13. Saragovi, H. U., Greene, M. I., Chrusciel, R. A., and Kahn, M. (1992). *Bio/technology*, **10**, 773.
14. Blankenmeyer-Menge, B. and Frank, R. (1990). In *Innovation and perspectives in solid phase synthesis* (ed. R. Epton), pp. 1–10. Chapman and Hall Medical, London.
15. Boldicke, T., Emery, S. C., Reubke, B., and Frank, R. (1991). In *Monoclonal antibodies in clinical oncology* (ed. A. A. Epenetos), pp. 1–10. Chapman and Hall Medical, London.
16. Geysen, H. M. (1986). *Mol. Immunol.*, **23**, 709.
17. Pinilla, C., Appel, J. R., Blanc, P., and Houghten, R. A. (1992). *Biotechniques*, **13**, 901.
18. Mutch. D. A., Rodda, S. J., Benstead, M., Valerio, R. M., and Geysen, H. M. (1991). *Peptide Res.*, **4**, 132
19. Bray, A. M., Maeji, M. J., Jhingran, A. G., and Valerio, R. M. (1991). *Tetrahedron Lett.*, **32**, 6163.

20. Maeji, N. J., Tribbick, G., Bray, A. M., and Geysen, H. M. (1992). *J. Immunol. Methods*, **146**, 83.
21. Van Bleek, G. M. and Nathenson, S. G. (1990). *Nature*, **348**, 213.
22. Conlan, J. W., Clarke, I. N., and Ward, M. E. (1988). *Mol. Microbiol.*, **2**, 673.
23. Miles, M. A., Wallace, G. R., and Clarke, J. L. (1989). *Parasitol. Today*, **5**, 397.
24. Champion, B. R., Page, K. R., Parish, N., Rayner, D. C., Dawe, K., Biswas-Hughes, G., Cooke, A., Geysen, M., and Roitt, I. M. (1991). *J. Ex. Med.*, **174**, 363
25. Burrows, S. R., Rodda, S. J., Suhrbier, A., Geysen, H. M., and Moss, D. J. (1992). *Eur. J. Immunol.*, **22**, 191.
26. Rodda, S. J., Benstead, M., and Geysen, H. M. (1993). In *Options for the control of influenza II* (ed. C. Hannoun *et al.*), pp. 237–43. Elsevier, Amsterdam.
27. Reece, J. C., Geysen, H. M., and Rodda, S. J. (1993). *J. Immunol.*, **151**, 6175.

8

Epitope mapping using libraries of random peptides displayed on phage

WILLIAM J. DOWER and STEVEN E. CWIRLA

1. Introduction

In 1985 Young *et al.* (1) introduced a method for cloning immunoreactive sequences, such as bacterial surface antigens, by expressing the proteins in λ plaques and identifying the clones of interest by probing with the appropriate antibodies. This method relied on the co-localization of the reactive peptide with the cloned DNA in a plaque or colony. This is a powerful technique, but it is difficult to screen more than a few million clones by this approach. Alternative methods, relying on the physical linkage of the protein and DNA, have recently become practical. These methods are known as peptide display techniques, and comprise the cloning of sequences that are expressed as peptides or proteins fused to a carrier protein and displayed on the outer surface of microorganisms. The surface peptide is immunoreactive, and the encoding DNA is carried within the microorganism. The clones of interest are isolated by affinity purification of the exposed peptides on a matrix of immobilized antibody or other binding protein. This approach, because it provides for the efficient isolation of individual microorganisms from a complex mixture, allows the screening of extraordinarily large numbers of clones.

The most advanced of these peptide display systems employs the fusion of peptides to the coat proteins of filamentous phage. This approach was devised by George Smith, who employed the minor coat protein pIII as the expression vehicle (2, 3). The method was initially conceived as a new type of expression cloning vector (4), but it has found its greatest use in the creation of very large libraries of randomized sequences; either completely random short peptides or randomized segments of larger peptides and proteins. For the purpose of this chapter, we will focus on the creation of libraries of peptides shorter than 20 residues made for the purpose of examining the binding specificity of the combining sites of antibodies—so-called 'epitope libraries'.

A practical obstacle to the creation of epitope libraries is the size of the libraries required. Coincident developments in the high-efficiency and high-capacity transformation of *Escherichia coli* brought the possibility of creating nearly complete hexapeptide libraries within reach (5). The construction and screening of large peptide libraries was accomplished and reported by three groups in 1990, demonstrating their use in identifying peptide ligands of antibodies (6, 7) and streptavidin (8). Filamentous phage display libraries of random peptides have since gained widespread use in the search for peptide ligands to antibodies, receptors, and enzymes.

2. Biology of the filamentous phage

Some understanding of the physiology of the filamentous phages is helpful in applying the phage-based systems most effectively to the screening of random peptide libraries, and we will cover some of the features most relevant to peptide display here. The life cycle of the filamentous phages is a fascinating subject in its own right, and an excellent review of this is contained in (9).

The infective cycle of the phage begins with its binding, via the minor coat protein pIII, to the pilus of the bacterial host cell. pIII is by far the largest protein (406 amino acids) on the virion, and by virtue of its adaptation to interacting with structures external to the phage, it is well suited to the display of foreign peptides. The C-terminus of pIII is embedded in the phage coat, and all of the four or five copies carried by the phage are attached in a small area at one end of the virion. The N-terminal domain of pIII has been shown to accept the fusion of small peptides with little detriment to its ability to bind the host cell (4). Upon binding to the pilus, the phage is drawn to the surface of the host cell, the coat proteins dissociate into the inner membrane, and the single-stranded loop of DNA that is the phage genome is taken into the cell and converted to the double-stranded, supercoiled replicative form (RF). Events to this point require only host functions and occur rapidly. The RF molecule serves as a template for transcription of phage proteins, some of which play a role in the generation of additional RF molecules by a rolling circle mode of plus-strand synthesis followed by complementary minus-strand synthesis, closing of the loop, and supercoiling. Among the phage gene products that accumulate are the five structural proteins destined for incorporation into newly formed phage particles. These are the major coat protein pVIII, the adsorption protein pIII, and the minor coat proteins pVI, pVII, and pIX—very small proteins found in fewer than 10 copies each on the ends of the virion.

pIII is synthesized as a pre-protein with an N-terminal 18-residue signal peptide which directs the assembly of pIII into the inner membrane. The signal peptide is then cleaved to release the mature N-terminus into the periplasm, as the C-terminus remains anchored in the inner membrane. As pIII awaits assembly into nascent phage particles, the fused peptides are exposed to the environment of the periplasm.

pVIII is produced in large amounts under control of a very strong promotor.

Wild-type phage of the f1, fd, M13 family contain about 2700 copies of pVIII assembled by interaction of their helical hydrophobic domains into the long thin tube characteristic of the filamentous phages. pVIII, like pIII, is made as a precursor protein with a 23-amino-acid signal peptide directing assembly of the protein into the inner membrane. After signal peptide cleavage, pVIII resides in the membrane with N-terminus toward the periplasm and C-terminus in the cytoplasm awaiting incorporation into nascent phage particles.

As RF replication proceeds, phage gene products accumulate. One of these, the cytoplasmic DNA binding protein pV, binds to newly synthesized plus-strands, preventing their conversion into double-stranded form and preserving the plus-strand for incorporation into phage particles. At some point the accumulation of pV effectively switches all phage replication into the production of plus-strand genomes. pV molecules coating the DNA are exchanged for pVIII as the DNA is extruded from the cell to produce the mature particles.

The trailing end of the phage DNA signals the addition of the several copies of pIII that decorate the end of the phage particle. The completed particle then exits the cell completely. Note that the filamentous phage do not lyse the cells as a means of dispersal. The cells remain chronically infected, growing slowly and producing phage particles for an indefinite period.

Any circular plasmid DNA that contains a filamentous replication origin and packaging signal has the potential to enter into a phage life cycle, producing infectious phage particles containing one strand of the plasmid DNA. The strand carrying the phage origin (plus-strand) is the strand which is packaged. Such episomes are called phagemids, replicating normally as plasmids under the control of host proteins, and entering into a mode of phage-like replication and packaging in the presence of certain phage proteins usually supplied by another phage genome (helper phage) introduced into the cell. To obtain preferential packaging of the phagemid genomes, the helper phage is modified to be somewhat debilitated, competing poorly with the phagemid for phage replication and packaging functions.

Phagemids used for peptide display purposes contain the gene for either pIII or pVIII with appropriate oligonucleotide cloning sites. These sequences are placed behind an inducible promotor to allow control of the timing and the level of expression of the coat protein–peptide fusion. The helper phage supply wild-type coat proteins, and the resulting phagemid particles display a mixture of wild-type and peptide-fused pIII or pVIII molecules.

3. An overview of the display systems

Filamentous phage have several properties making them attractive for use as peptide display vectors. They are very well characterized, easy to work with, and the surface of the virion is of low complexity.

We routinely employ three variations of the phage display systems, and a similar method based on a non-phage vector. The original phage–pIII system contains the cloned oligonucleotide inserted near the 5'-end of gene III in the phage genome. Peptides encoded by the oligonucleotides are expressed fused at or near the N-terminus of the minor coat protein pIII, and are assembled into the surface of the phage particles where they are available for binding to antibodies. Each of the four or five copies of pIII on a phage carries a peptide. The vectors we use are derived from a tetracycline resistance transducing strain of phage fd, fd-tet (10). Vectors derived from other strains of filamentous phage such as M13 are also in use. A schematic diagram of the display and screening arrangements is shown in *Figure 1*.

A variation of the pIII display system employs a phagemid vector. A number of these phagemid vectors have been constructed for various purposes. These vectors contain gene III with oligonucleotide cloning sites. Usually the pIII promotor is replaced with an inducible promotor to permit controlled expression of the fusion protein. Copies of wild-type pIII, along with other required phage gene products, are supplied by a 'helper' phage, and the number of peptides displayed on each phagemid particle may be varied from zero to five by control of the relative numbers of wild-type and peptide-bearing pIII molecules expressed in the cell. The phagemid DNA is packaged in preference to the helper DNA, providing the linkage between cloned DNA and displayed peptide.

Another structural protein, the major coat protein pVIII, has also been exploited as a display vehicle. While peptides may be fused to each of the several thousand copies of pVIII that coat the phage, the size of the peptides in this arrangement is reportedly limited to six or fewer residues (11). Larger peptides may be displayed if the phage are coated with a mixture of wild-type and fused pVIII, and this is most conveniently done with a phagemid vector (11, 12). As with the pIII vectors, the coat protein containing the foreign peptide is expressed under control of an inducible promotor permitting adjustment of the ratio of wild-type (supplied by the helper phage) and fused pVIII molecules. The phagemid vector we have constructed, pAFF8, provides expression of the fusion protein under the control of the arabinose promotor resulting in the display of several hundred copies of the peptides on the surface of each phage (E. A. Peters *et al.* unpublished data).

We also employ a non-phage display system that creates a fusion of peptides to a DNA binding protein to obtain the physical linkage between peptide and DNA (13). LacI carries the peptide fused to its C-terminus. The plasmid encoding the LacI–peptide fusion contains two copies of the LacO sequence, the natural binding site of LacI. Because of their compartmentalization within each cell, the LacI–peptide fusion molecules bind to LacO sites on the plasmids which encode them. The resulting complex is sufficiently stable to undergo an affinity purification procedure on immobilized target binding proteins done under conditions similar to the screening of peptides on

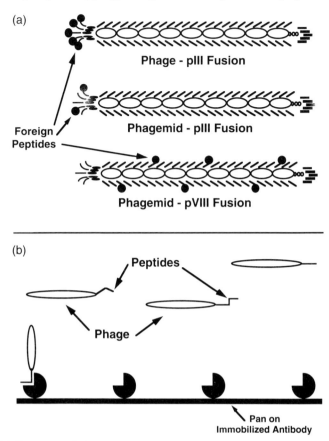

Figure 1. (a) Structure of phage display systems. Peptides are displayed fused to a coat protein of filamentous phage. Random oligonucleotides are cloned into the 5'-region of the gene for either the adsorption protein pIII, or the major coat protein pVIII. The episome encoding the peptide–protein fusion is either a phage genome, serving as the only source of the coat protein; or a phagemid, which supplies only a portion of the coat protein, the remainder (as wild-type) supplied by a helper phage. With phage vectors, all copies of the coat protein may carry a peptide; with phagemid vectors, a mixture of peptide-fused and wild-type coat protein are incorporated into the virions. (b) Phage library screening. Phage particles carrying ligand peptides are selected by exposing the entire library of phage clones to immobilized antibodies, washing away the unbound phage, eluting and amplifying the bound phage, and re-exposing those to the antibody matrix. The process is repeated several times to enrich for those phage-carrying peptides that bind to the antibody. The sequence of the active peptides is deduced from the DNA sequence in the oligonucleotide cloning site of the phage genome.

phage. Mention of this system is included here because the collections of peptides expressed in this way are somewhat different from those displayed by the phage systems. In particular, the peptides are expressed with a free C-terminus rather than the free N-terminal display typical of expression on the phage coat proteins. Thus, if a free C-terminus is favoured for recognition by the antibody under study, a LacI library is more likely to contain the required peptides. The LacI system also produces peptides that have been subjected to a set of biases different from those surviving the phage expression. The LacI fusions are strictly cytoplasmic, while the pIII and pVIII molecules are assembled in the inner membrane, and the peptides they carry reside for a time in the periplasm prior to incorporation into phage. The LacI fusions are less subject to degradation by the subset of proteases and peptidases restricted to the periplasm, and are not affected by membrane transport and processing specificities that may limit the diversity of peptides displayed on pIII and pVIII. Protocols for the use of the LacI fusion system are not included in this chapter, but the reader is referred to (13) for a detailed description of this system.

4. Detailed description of each phage display system

The phage pIII system is the simplest of the systems to use, avoiding the steps of pIII induction and helper phage rescue necessary with phagemid vectors. We developed a vector for this purpose, derived from fd-tet, which we call fAFF1 (*filamentous AFFinity* phage) (7). Reasonably large libraries can be conveniently created in this vector. We routinely construct libraries of 10^8 to greater than 10^9 members displaying peptides of 6 to 20 residues (14). fAFF1, because it is a substitution vector, permits the peptides to be displayed at the very N-terminus of pIII; thus, the first residue of the foreign peptide is the first residue of the mature fusion protein. This arrangement places a variable residue at the C-terminal side of the signal peptidase cleavage site and initially raised some concern about the effect on processing adjacent to certain amino acids. Our analyses of several libraries have shown a close-to-expected frequency of occurrence of codons for most amino acids at all of the variable positions in clones that actively express mature phage particles, although at least one amino acid, proline, appears to be excluded from the $P_{1'}$ position (the N-terminus of the mature fusion protein). Other peptide-specific biases include effects on assembly of the peptide–pIII fusion into the inner membrane. About 5% of clones show reduced yield of mature phage. Most of these clones have multiple basic residues in the early mature region, a composition that we have shown to interfere with *sec*-dependent transport of pIII (E. A. Peters *et al.* in preparation). These known biases are of minor significance for the diversity of peptides available in the libraries. There are, however, other potential biases, especially those caused by exposure of the peptides to peptidases and proteases in the cells, that have not yet been systematically studied.

In fAFF-infected cells, the peptide–pIII fusion protein is the only source of pIII; therefore, all copies of pIII on each phage particle will carry a foreign peptide. This allows multivalent interaction of the phage with immobilized binding sites, and results in an avid binding of rather weak ligands even under conditions of vigorous and extended washing (7, 15). Our experience with antibodies, as well as receptors, indicates that peptides with affinities as weak as 50 to 100 μM can be recovered with the pIII phage system (R. W. Barrett and W. J. Dower unpublished).

A disadvantage of the phage system is its relative inconvenience, compared with the pIII phagemid system, in constructing very large libraries. The limiting material in library construction is vector DNA, and the yield of phage RF DNA is rather low, typically 200 to 400 μg/litre. Losses from subsequent preparation steps are often as much as 50%, providing a final yield of *prepared* vector DNA of ~100 μg/litre of culture. The electrotransformation of the large (9.2 kb) phage RF is also low when compared with phagemid DNA, the number of transformants produced is about 10^9 per μg, and drops to about 10^7 per μg for vector ligated to oligonucleotides.

The phagemid–pIII system requires more effort during the screening process, but has two advantages over the phage–pIII system. Yields of double-stranded phagemid DNA are five- to ten-fold higher than phage RF, and the transformation efficiencies are several-fold higher, permitting a library of given size to be constructed with less effort in phagemids than in phage. In addition, because the pIII–peptide fusion is expressed under the control of an inducible promotor (arabinose in our pAFF (*p*hagemid *AFF*inity vectors) series of vectors) the number of pIII molecules that carry a peptide can be adjusted. This permits switching between the display of several copies (up to five) to obtain the sensitive, multivalent attachment to immobilized binding protein, to as few as one peptide per phage particle, a configuration designed to selectively enrich for peptides with higher intrinsic affinities for the binding sites.

The pVIII display system differs from the pIII systems in several important ways. We employ phagemid vectors, a series called pAFF8, for pVIII display. We estimate that there are roughly 2000 copies of pVIII on the mature pAFF8 particle. Expression of the pVIII–peptide fusion is driven by the arabinose promotor and can be controlled to produce very few peptides under conditions of glucose repression, or up to 500 copies per phage particle under conditions of strong induction (E. A. Peters and W. J. Dower unpublished). Displaying hundreds of peptides per phage provides a large number of attachment points and a very high avidity for the immobilized binding site; we have recovered peptide ligands with affinities as low as 500 μM from complex pVIII libraries (R. W. Barrett and W. J. Dower unpublished).

5. Introduction to the protocols

We have provided detailed protocols for: (i) the construction of peptide display libraries in the phage vector fAFF1 (*Protocols 1–4*); (ii) the screening of phage libraries (*Protocols 5–7*) and (iii) a series of auxiliary protocols useful in characterizing the libraries and the individual phage clones recovered in screening target antibodies (*Protocols 8–11*). In the text accompanying the protocols, we discuss the important issues and attempt to explain the reasons behind many of the steps of the procedures.

6. Constructing peptide display libraries in filamentous phage vectors

Protocols 1–4 contain a description of the procedure from preparation of the vector DNA and the oligonucleotides through the amplification and harvesting of the phage constituting the libraries.

We generally prepare and characterize vector DNA in batches of 2 to 4 litres of culture (*Protocol 1*). Our primary phage vector, fAFF1, is produced in modest yields of 200 to 400 μg/litre. Good phage production is dependent on efficient aeration of the cultures, and the cells are grown in large flasks with vigorous shaking. The cloning scheme we employ for both the pIII and pVIII systems consists of the removal of a short stuffer fragment that spans the sequence encoding the signal peptidase cleavage site in the pIII or pVIII precursors. This exposes a pair of *Bst*XI sites which, because of the internal degeneracy of the recognition sites for *Bst*XI, have 3'-overhangs chosen to be neither self-complementary nor complementary to one another. This arrangement allows oriented cloning of the oligonucleotides and a low background of insert-lacking clones. The removal of the stuffer fragment following *Bst*XI digestion is accomplished by centrifuging the digested vector down a step gradient of potassium acetate. The typical yield for this step is only about 50%. The overall yield of prepared vector DNA is 100 to 200 μg/litre of culture. fAFF1 RF DNA is the limiting material in library construction, and we often use 100 μg to prepare each primary library, with the expectation of obtaining about 10^9 transformants.

Once the vector DNA has been purified free of the stuffer fragment, the oligonucleotide is prepared and ligated (*Protocols 2* and *3*) as an annealed complex of three pieces (*Figure 2*):

(a) a long oligonucleotide containing a 3'-terminus complementary to the second *Bst*XI site, a spacer region encoding the linker from the coat protein to the variable peptide, the variable oligocodon region, and a 5' segment encoding the reconstruction of the C-terminal portion of the signal peptide

(b) a small 'half-site oligonucleotide' that anneals to the 5'-fixed region of oligonucleotide (ON-1) to create the 3' overhang complementary to the first *Bst*XI site and

(c) a second half-site oligonucleotide that anneals to the 3' end of oligonucleotide (ON-1) to provide an additional ligation point to the second *Bst*XI site (see *Figure 2* and (7)).

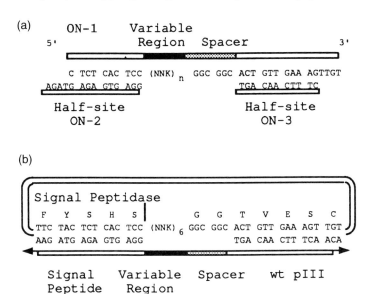

Figure 2. (a) Structure of the degenerate oligonucleotides (ON) for library construction in the pIII phage vector fAFF1. (b) Sequence of the reconstructed pIII gene in the vicinity of the cloning site after insertion of the degenerate oligonucleotide.

The variable oligocodon of oligonucleotide (ON-1) is created during synthesis with the codon motif $5'$-$(NNK)_n$-$3'$, where N is an equimolar mixture of phosphoramidite-activated bases A, C, G, and T; and K is an equimolar mixture of G and T. This motif may be obtained either by programming the synthesizer to deliver equal amounts of each base at each cycle, or by mixing the desired ratio of bases in a single bottle for delivery at each cycle. The codon motifs NNK and NNS (where S is an equimolar mixture of G and C) encode 32 codons, representing all 20 genetically coded amino acids: a single codon for 12 amino acids, two codons for each of five amino acids, three codons for three amino acids, and a single (amber) stop codon.

The oligonucleotides are purified, phosphorylated, annealed, and ligated to the prepared vector DNA. After ligation, the remaining gap is filled in with polymerase. This last step reduces the occurrence of aberrant sequences (deletions, etc.) recovered from the libraries.

For primary library construction, we utilize *E. coli* host strains with the highest transformation efficiencies. These tend to be Rec+ strains, probably because of their higher viability. Some of the best strains for transformation are derived from MC1061. Mutant host strains are useful for particular purposes. For example, cells carrying amber suppressors allow the creation of libraries with longer random peptides without interruption by the amber stop codons encoded by the NNK motif. Particular peptide sequences can be detrimental to the expression of the phage. We have shown that multiple basic residues in the early mature region of pIII can cause greatly decreased production of phage particles; a defect that is largely overcome when the phage are grown in strains carrying the secretory mutation *prl*A (E. A. Peters *et al.* in preparation). It seems likely that certain of the peptides that would occur in a random library are vulnerable to degradation by cellular proteases and peptidases, and the use of protease mutants as hosts for library construction may be helpful in increasing the diversity of sequences expressed in the libraries.

After ligation of the annealed oligonucleotides to the vector DNA, the DNA is precipitated with ethanol, resuspended in buffer, and transformed into cells made competent for electroporation (5). Ligation reactions contain mild inhibitors of electroporation and the precipitation step improves by several fold the transformability of the DNA. Two characteristics of electroporation make possible the construction of the extremely large libraries required for random library screening. First, the transformation efficiencies are typically about ten-fold higher than the best chemical transformation protocols provide; second, and more importantly, electrotransformation has a capacity for added DNA that is several orders of magnitude greater than that of chemical transformation (5, 14). Each preparation of competent cells should be tested for transforming efficiency prior to committing the library DNA. Electrocompetent MC1061 cells should produce $(2-5) \times 10^{10}$ transformants per microgram with a clean and well calibrated stock of pUC DNA. Construction of very large libraries requires prodigious amounts of DNA, and test transformations should first be performed with small amounts of the ligated DNA to verify adequate transformation efficiency before commiting the whole of the ligated DNA. After transformation of the products of ligation followed by non-selective outgrowth, a small aliquot of the transformation is plated to determine the number of independent recombinants in the library. The infectivity of several of the clones should also be determined as described in *Protocol 8*, and should be within about 10% of the expected value $((1-s)^n$, where s is the expected frequency of stop codon appearance at each position (1/32 for NNK or NNS motifs) and n is the length of the peptides encoded in the library). To check the quality of the library, a number of individual clones, perhaps 20, should be sequenced and inspected for anomalies.

The large-scale transformations are pooled after 1 h of outgrowth, and inoculated into large cultures containing antibiotic. A rule of thumb is to culture about $(3-5) \times 10^8$ transformants per litre. The procedure for amplifying the libraries is described in *Protocol 3*. Each host cell produces hundreds of phage particles during each generation. The burden of this production slows the growth of the cells, resulting in a population doubling time of 1 to 1.5 h. The purpose of the amplification step is to create many copies of each phage; but the more a population is amplified, the greater the likelihood of introducing bias into the population. Therefore we generally choose a conservative level of amplification. The protocol described here yields about 10^5 infective phage per transformant. The cells expand about 1000-fold (10 doublings) in this time. After harvest of the expanded phage population (*Protocol 4*), aliquots of about 1000 'library equivalents' are prepared and stored at -20 or $-70\,°C$. One library eqivalent (LE) is a number of phage particles equal to the number of independent transformants that constitute the library. Each aliquot of 1000 LE is adequate for a single screening round.

Protocol 1. Phage library construction: vector preparation

Reagents

- DH5αfAFF1 phage (may be obtained from Affymax Research Institute).
- *Bst*XI (New England Biolabs)
- LB-tet medium (1% Bacto tryptone, 0.5% Bacto yeast extract, 0.5% NaCl, 20 μg/ml tetracycline)
- 10× NE buffer 3 (0.5 M Tris–HCl, pH 7.9, at 25°C, 1 M NaCl, 100 mM MgCl$_2$, 10 mM dithiothreitol)
- Phenol:chloroform (1:1)
- TE buffer (10 mM Tris–HCl, 1 mM EDTA, pH 8.0)
- Potassium acetate solutions (5, 10, 15, and 20% w/v in 2 mM EDTA, 1 μg/ml ethidium bromide)
- Reagents for isolating fAFF1 RF DNA (see (16))
- 7.5 M ammonium acetate

A. *fAFF1 replicative form DNA isolation*

1. Inoculate 5 ml of LB-tet medium with a single colony of DH5αfAFF1. Grow at 37°C with shaking overnight.

2. Add the 5 ml of overnight culture to a 2.8 litre Fernbach flask containing 1 litre LB-tet medium. Grow at 37°C shaking vigorously overnight.

3. Isolate fAFF1 RF DNA using the alkaline lysis method followed by CsCl / ethidium bromide equilibrium centrifugation. (Detailed protocols are found in (16), pp. 1.38–1.46.) The expected yield of fAFF1 RF DNA is 200–400 μg/litre.

B. Bst*XI digestion of fAFF1 DNA*

1. Combine 200 μg of fAFF1 RF DNA with 40 μl 10 × NE buffer 3 and 20 μl *Bst*XI (200 units) in a total volume of 400 μl. Incubate the reaction mixture overnight at 55°C in a covered water bath.

2. Extract the reaction mixture twice with an equal volume of phenol/chloroform and once with chloroform. Divide the aqueous phase from the final extraction into two microcentrifuge tubes.

3. Add one-half volume 7.5 M ammonium acetate and two volumes ethanol. Spin in a microcentrifuge for 1 min at room temperature.[a]

4. Resuspend the DNA pellets in 200 μl TE buffer and repeat step **3**.

5. Wash pellets with 200 μl 70% ethanol, spin for 1 min in a microcentrifuge and resuspend pellets in 200 μl TE buffer. Combine into one tube and determine the DNA concentration by measuring absorbance at 260 nm.

Protocol 1. *Continued*

C. *Potassium acetate step gradient centrifugation*[b]

1. Two gradients are sufficient to purify 200 μg of *Bst*XI digested fAFF1 DNA. For each gradient add 1 ml of 5% potassium acetate solution to a 13 × 55 mm polyallomer ultracentrifuge tube. Slowly underlay 1 ml of each of the potassium acetate solutions in order of increasing density using a 1 ml syringe.

2. Carefully layer 100 μg of *Bst*XI-digested fAFF1 DNA in 0.5 ml TE buffer on top of the gradient. Centrifuge at 280 000 *g* (48 000 r.p.m. in a SW50.1 rotor) for 3 h at 20 °C.

3. Visualize the broad band of linearized DNA that is contained in the lower part of the tube using a longwave hand-held UV lamp. Carefully remove the band with a syringe and extract the solution several times with water-saturated butanol to remove the ethidium bromide.

4. Precipitate the DNA by adding 0.1 volume 3 M sodium acetate and 2 volumes ethanol. Wash the pellet with 70% ethanol and resuspend in TE buffer. Determine the DNA concentration by measuring absorbance at 260 nm.

[a] Several precipitations in the presence of ammonium acetate will not remove all of the stuffer fragment; however, vector DNA that is prepared in this manner is adequate for library construction. The stuffer fragment can be completely removed from the precipitated DNA by potassium acetate step gradient centrifugation.
[b] Modification of protocol described by Arufo and Seed (17).

Protocol 2. Phage library construction: oligonucleotide preparation

Equipment and reagents

- T4 kinase (New England Biolabs)
- OPC columns (Applied Biosystems)
- Fluorescent aluminium oxide TLC plate (J. T. Baker)
- Sephadex G-25 spin column (Pharmacia)
- 8% polyacrylamide/7M urea/TBE gel

- 1 M triethylamine bicarbonate (TEAB, adjust to pH 7.0 with CO_2)
- 10× kinase buffer (1 M Tris–HCl, pH 7.5, 100 mM $MgCl_2$, 50 mM dithiothreitol)
- 10 mM ATP

A. *Synthesis*

1. Synthesize oligonucleotides (ON) with the following sequences (see *Figure 2*):

   ON-1: 5′-<u>CTCTCACTCC</u> (NNK)*n* <u>ACTGTTGAAAGTTGT</u>-3′

N = A,C,G,T (equimolar), K = G,T (equimolar)

n is the desired number of residues in the random peptides

ON-2: 5'-GGAGTGAGAGAGTAGA-3'

ON-3: 5'-CTTTCAACAGT-3'

Synthesize ON-1 with trityl-off, deprotect, and purify on a denaturing polyacrylamide gel as described below. Synthesize ON-2 and ON-3 with trityl-on, deprotect and purify on OPC columns according to the manufacturer's directions (Applied Biosystems). Phosphorylate purified oligonucleotides with T4 kinase (see below).[a]

B. *Gel purification*

1. Elute oligonucleotide from the synthesis column with 2 ml of 30% NH$_4$OH (30 min at room temperature) and deprotect overnight at 55 °C. Lyophilize the crude deprotected oligonucleotide and resuspend in 400 μl water.

2. Use an 8% polyacrylamide/7 M urea/TBE gel to purify oligonucleotides from 50–100 bases long. As much as 500 μg of crude oligonucleotide can be loaded into a single 2 cm well of a 24 × 15 cm gel with 1.5 mm spacers. (A detailed protocol for preparing and running the gel can be found in (18), pp. 2.12.1 – 2.12.5.)

3. Following electrophoresis, transfer the gel to plastic wrap and place on top of a fluorescent TLC plate. Visualize the dark oligonucleotide band by briefly shadowing with a shortwave hand-held UV lamp (to prevent radiation damage to the oligonucleotide, expose the gel to UV for as short a period as needed to mark the band). Cut out the band, transfer it to a microfuge tube and crush it into small pieces. Add 1 ml of 10 mM TEAB (made from a 1M stock) and elute the DNA overnight at room temperature.

4. Centrifuge briefly to pellet the polyacrylamide gel fragments. Transfer the oligonucleotide-containing solution to a microfuge tube, reduce volume to about 200 μl in a Speedvac and load onto a 1.5 ml Sephadex G-25 spin column. Collect the flow-through material, dry down the sample in a Speedvac, and resuspend the pellet in water. Measure absorbance at 260 nm to estimate the DNA concentration.

C. *Kinase reaction*

1. Combine the following in a microcentrifuge tube:
   - 5 μg purified oligonucleotide
   - 2 μl 10× kinase buffer
   - 2 μl 10 mM ATP
   - 2 μl T4 kinase (20 units)
   - water to 20 μl total reaction volume

Protocol 2. *Continued*

2. Incubate at 37 °C for 90 min and stop the reaction by heating to 70 °C for 10 min. Store the phosphorylated oligonucleotide at −20 °C.

[a] Alternatively, all three oligonucleotides may be chemically phosphorylated during synthesis using the 5′ phosphate-ON reagent that is available from Clontech. Deprotect oligonucleotides and gel purify.

Protocol 3. Phage library construction: ligation of oligonucleotides with *Bst*XI-digested fAFF1 DNA [a]

Equipment and reagents

- *E. coli* MC1061 (American Type Culture Collection)
- T4 DNA ligase (New England Biolabs)
- Sequenase 2.0 (United States Biochemicals)
- Electroporation apparatus (Bio-Rad Gene Pulser)
- 10 × annealing buffer (200 mM Tris–HCl, pH 7.5, 20 mM MgCl$_2$, 500 mM NaCl)
- 10 × ligation/synthesis buffer (100 mM Tris–HCl, pH 7.5, 50 mM MgCl$_2$, 20 mM dithiothreitol, 10 mM ATP)
- dNTP mix (25 mM of each dNTP)
- TE buffer (*Protocol 1*)
- SOC medium (2% Bacto tryptone, 0.5% Bacto yeast extract, 10 mM NaCl, 2.5 mM KCl, 10 mM MgCl$_2$, 10 mM MgSO$_4$, 20 mM glucose)
- LB-tet medium and agar plates (*Protocol 1*)
- 3 M sodium acetate

A. *Annealing*

1. Combine the following in a 1.5 ml microcentrifuge tube:
   - 40 µg fAFF1 (6.6 pmol, *Bst*XI-digested and purified)
   - X ng ON-1 (16.5 pmol) [b]
   - 75 ng ON-2 (16.5 pmol)
   - 1.2 µg ON-3 (330 pmol)
   - 40 µl, 10 × annealing buffer
   - water to 400 µl total reaction volume

2. Incubate at 70 °C for 5 min and then cool slowly to room temperature to allow oligonucleotides to anneal.

B. *Ligation and fill-in reaction*

1. Add 45 µl 10× ligation/synthesis buffer, 5 µl T4 DNA ligase (2000 units) and incubate overnight at 14 °C.

2. Fill in the single-stranded gap by adding 5 µl of dNTP mix and 1 µl of Sequenase 2.0 (13 units). Incubate at 37 °C for 90 min.

3. Precipitate DNA by adding 0.1 volume 3 M sodium acetate and 2 volumes ethanol. Wash the pellet with 70% ethanol, dry, and resuspend in 40 µl TE buffer.

C. *Electrotransformation*

1. Prepare electrocompetent *E. coli* MC1061 cells as described in (5). A test transformation with MC1061 should yield $(2-5) \times 10^{10}$ transformants per microgram of pUC DNA.

2. Transform the ligated DNA by electroporation into MC1061. Perform eight electrotransformations, each containing 100 μl of cells and 5 μg of DNA, by pulsing at 12.5 kV/cm for 5 msec (as described in (5)).

3. Following the pulse, *immediately* transfer the cells from each transformation to 2 ml SOC medium and incubate without selection for 1 h at 37 °C.

4. Pool the transformations and plate a small aliquot on LB-tet agar plates to assess the transformation efficiency. Add the remainder to 1 litre of LB-tet medium and grow through approximately 10 doublings at 37 °C to amplify the library.

[a] This protocol is designed for 40 μg of fAFF1 DNA and an expected yield of $(1-5) \times 10^8$ recombinants. The typical transformation efficiency of ligated fAFF1 DNA is $(0.5-2) \times 10^7$ transformants per microgram of vector. The transformation efficiency of uncut fAFF1 DNA is about 2×10^9 per μg using electrocompetent MC1061 cells. Transform the electrocompetent cells with uncut plasmid DNA to check the efficiency prior to using them for library construction.

[b] The mass of ON-1 to be used depends on the size of the oligonucleotide. The molar ratio of fAFF1: ON-1: ON-2: ON-3 is 1 : 2.5 : 2.5 : 50.

Protocol 4. Phage library construction: harvest of amplified phage library

Reagents

- 20% polyethylene glycol (PEG, average molecular weight 8000)/2.5 M NaCl
- PBS (0.137 M NaCl, 4.3 mM Na_2HPO_4, 1.4 mM KH_2PO_4, 2.7 mM KCl, pH 7.4 at 25 °C)

A. *Isolation and storage of phage*

1. Centrifuge the 1 litre overnight culture at $12\,000g$ (8000 r.p.m. in a JA-10 rotor) for 15 min at 4 °C. Transfer the supernatant to a new bottle and repeat the spin.

2. Transfer the cleared supernatant to a new bottle and precipitate the phage by adding 0.2 volumes of PEG/NaCl. *Mix well* and hold on ice for 1 h.

3. Pellet phage by centrifugation at $12\,000\ g$ for 15 min. Remove as much of the supernatant as possible and resuspend the phage pellet in 20 ml of PBS. Heat the phage solution to 70 °C for 15 min to kill any residual bacteria. Titre phage stock (as described in *Protocol 8*); the suspension may be stored for up to 24 h at 4 °C while the titre is determined.

Protocol 4. *Continued*

4. Soon after determining the titre, dispense the phage into aliquots of 1000 LE for long-term storage at $-20\,°C$.

B. *Library analysis*

1. Assess the cloning efficiency by checking the infectivity of phage produced from individual transformants. Infectivity is determined by titreing phage as described in *Protocol 8*. fAFF1 contains a -1 frameshift mutation in pIII that results in non-infective phage particles ($<10^3$ TU/ml). By inserting an oligonucleotide the correct translation frame is restored to pIII to produce infective phage ($\sim 10^{10}$) TU/ml).

2. Pick about 10 or 20 random transformants, prepare phage and single-stranded DNA from each clone, and sequence the DNA in the variable region (see *Protocol 11*). These sequences are analysed to assess the frequency of anomalous clones in the library (deletions, etc.).

7. Screening the libraries

There are two basic strategies in use for selecting phage displaying peptide epitopes. One strategy employs an affinity matrix of immobilized antibodies to which the phage library is exposed, capturing the active phage directly on the matrix. An alternative approach entails bringing the phage and antibody together in solution, and then recovering the phage–antibody complexes by capturing the antibody (3). Both methods have theoretical and practical benefits. We utilize almost exclusively the immobilized target method, and the protocols in this chapter are devoted to this approach.

7.1 Immobilization and characterization of the target antibody

A simple and usually effective method of immobilizing antibodies in preparation for screening is adsorption directly to the wells of microtitre plates designed for this purpose. We have used plates from a variety of commercial suppliers with good results. The conditions appropriate for adsorbing a particular antibody must be determined empirically by testing different brands of plates, concentrations of antibody applied, adsorption buffers (with particular attention to pH), and the time and temperature of incubation. *Protocol 5* describes conditions that we have found effective for immobilizing a variety of antibodies from several species, and provides a good starting point for customizing the process for any target antibody. Another strategy for immobilizing antibodies may be employed, whereby a second, polyclonal antibody, goat anti-mouse IgG, for example, is adsorbed to the wells and the primary target is then captured. This format is particularly convenient if a large number of different target antibodies are to be screened, but the presence of the second antibody contributes a variety of binding sites and can complicate the interpretation of screening results.

As part of the preparation for screening, it is necessary to demonstrate that the antibody retains its binding activity once immobilized. This is done with simple binding experiments employing a known ligand. One must either label the ligand, or have available another method for detecting the retention of the ligand on the immobilized antibody. This can be done with a non-blocking antibody against the ligand, for example. This binding assay is also used to determine the conditions for screening the antibody of interest. In certain circumstances, a ligand for the target antibody will be unknown or unavailable. In this situation, one can only determine that the antibody has been adsorbed to the surface (by detection with a labelled second antibody) and proceed with screening in the hope that a significant portion of the immobilized target molecules present an active combining site. The lack of a known ligand also complicates the determination of specificity of peptides subsequently recovered from the library. After adsorption of the antibody, the plates are blocked with bovine serum albumin (BSA), gelatin, or another inert protein. When the wells are properly blocked, non-specifically bound phage are usually recovered at a level of 10^{-4} to 10^{-6} of input.

Magnetic beads are also useful as an immobilization substrate. Beads with a variety of surface derivatizations are available. Antibodies can be adsorbed directly, captured on a second antibody, covalently coupled through various linkages, biotinylated and captured on streptavidin-coated beads, and so on.

Other derivatized supports such as chromatography resins may also be used to create the antibody matrix. We have little experience with the use of these for screening phage libraries. The non-specific binding of phage to some types of supports may be high and should be checked before attempting to screen a library.

7.2 Overview of the screening process

To the matrix of active immobilized antibody, we apply about 100 to 1000 LE of phage to the wells of an ELISA microplate. A typical library might have about 3×10^8 independent transformants; a thousand equivalents, about 3×10^{11} transducing units (TU, i.e. infective phage), would be used in the first round of screening. We have not determined systematically that this excess of phage is necessary (or, for that matter, sufficient). Fewer LE may work well for a given antibody. In the initial round, phage are applied at a concentration of about 10^{11} TU in 100 µl per well. The incubations and the washes are carried out in the cold to minimize degradation of the peptides by proteases that may be present, and to decrease the dissociation of phage from the surface once bound.

Bound phage are eluted with acid (pH 2.2) for 10 min at room temperature (*Protocol 6*). Other elution protocols may be employed alternatively, in parallel, or sequentially to recover phage that might resist elution with acid. These protocols include elution with base (pH 11) for 10 min at room temperature, or high temperature (70°C) for 10 min. The phage are reasonably stable to these treatments.

The pooled, neutralized phage are then amplified in preparation for the next round of screening (*Protocol 7*). We typically employ *E. coli* strain K91 as a host for this step, but special mutants may be desirable for certain purposes. The chosen strain must express an F-pillus to render it infectable by filamentous phage. The host cells are prepared by growing to mid-log phase and concentrating them. The phage are added at a multiplicity of infection of 10^{-2} to 10^{-3} and left undisturbed in contact with the cells for 20 to 30 min at 37°C. Shaking is avoided as it may disrupt the association of the phage with the cells during the early stages of infection. We grow the infected cells on large LB-tet plates to amplify phage between rounds as described in *Protocol 7*. After overnight growth, the amplified phage are harvested, concentrated, and rescreened on freshly immobilized target antibody. Recoveries are tracked by titreing before and after amplification. To reduce the background, about tenfold fewer phage are added in each subsequent round of screening. There are several tests that can be done after three or four rounds to check for enrichment of ligand clones. Finally, the recovered clones are analysed to determine their specificity for the target antibody.

Protocol 5. Phage library screening: antibody immobilization

Equipment and reagents

- ELISA microplates (Beckman Instruments or Dynatech Immulon-4)
- PBS (*Protocol 4*)
- PBS/1% BSA

Method

1. Dilute antibody to 100 μg/ml[a] in PBS. Add 50 μl to six wells of a 96-well ELISA microplate and incubate at 37 °C for 1 h.

2. Wash the plate with PBS. Completely fill wells with PBS/1.0% BSA and incubate for 1 h at 37 °C to block the plate.

3. Wash the plate with PBS and store covered for up to 24 h at 4 °C until ready to screen the phage library.

[a] The amount of antibody that binds to the plate may vary, therefore some experimentation is required to determine a reasonable coating concentration. One should also assay the activity of the immobilized antibody prior to panning. Alternative methods of immobilization using magnetic beads or secondary antibodies have also been used successfully, but do not seem to offer any advantage over directly coated plates.

Protocol 6. Phage library screening: panning

Reagents

- PBS (*Protocol 4*)
- PBS/0.1 BSA
- 2M Tris
- Elution buffer (0.1 M HCl (pH 2.2 with glycine)/0.1% BSA)

A. *Adsorption of phage*

1. Add about 1000 LE for the first round of panning. Dilute the phage library in 600 μl PBS/0.1% BSA and add 100 μl to each of the antibody-coated wells.[a]

2. Incubate the plate for 2 h at 4 °C. Overnight incubation at 4 °C does not seem to improve the recovery of specific phage clones.

3. Wash the plate by filling the wells with cold PBS or PBS/0.1% BSA and aspirating approximately five times.

4. Fill the wells with PBS or PBS/0.1% BSA and incubate at 4 °C for 30 min. Repeat step **3**.

B. *Elution*

1. Add 100 μl of elution buffer to each well. Incubate at room temperature for 10 min.

2. Transfer eluates (600 μl total) to a microfuge tube and neutralize with 35 μl of 2 M Tris base (pH unadjusted).

3. Titre the phage eluate to determine the number of phage recovered and amplify as described in *Protocol 7*.[b]

---

[a] BSA is included in the binding buffer to reduce the number of phage that might bind specifically to the immobilized BSA used to block the plate.
[b] 10^{10} to 10^{11} TU of the amplified phage stock is typically used for the next round of panning.

Protocol 7. Phage library screening: amplification between rounds

Reagents

- *E. coli* K91 (American Type Culture Collection)
- LB medium (1% Bacto tryptone, 0.5% Bacto yeast extract, 0.5% NaCl)
- LB-tet agar plates (LB medium, 20 mg/l tetracycline, 15 g/l Bacto agar)
- PEG/NaCl (*Protocol 4*)

Protocol 7. *Continued*

A. *Infection and growth*

1. Inoculate 5 ml of LB medium with a single colony of *E. coli* K91. Grow at 37 °C with shaking overnight.

2. Add 1 ml of the overnight culture to a 100 ml flask containing 20 ml LB medium and grow with shaking to A_{600} = 0.2–0.4. Pellet cells by centrifugation and resuspend in 2 ml of LB medium.

3. Add 600 µl of the concentrated cells to 600 µl of phage eluate. Incubate at 37 °C for 20 min with no shaking.

4. Plate 400 µl of infected cells on each of three 150 mm LB-tet agar plates. Incubate plates overnight at 37 °C.

5. The next day the plates should be covered with a lawn of bacterial cells. Add 10 ml of LB medium to each plate, let it stand for 10 min and gently scrape with a sterile spreader to recover bacterial cells and phage. Isolate phage as described below.

B. *Phage isolation*

1. Centrifuge the cell suspension at 12 000 *g* for 15 min.

2. Transfer the cleared supernatant to a new bottle and precipitate phage by adding 0.2 volumes of PEG/NaCl. Mix well and let stand on ice for 1 h.

3. Pellet phage by centrifugation at 12 000 *g* for 15 min. Remove as much of the supernatant as possible and resuspend the phage pellet in PBS. Heat the phage suspension to 70 °C for 15 min to kill any residual bacteria. Titre the phage stock (*Protocol 8*) and store the suspension for up to 24 h at 4 °C or indefinitely at −20 °C.

8. Characterization of the recovered clones

The issue of specificity of binding is a critical one, and especially so for those sequences bearing no recognizable resemblance to the antigen and thought to mimic discontinuous epitopes. Phage recovered in each round of screening may bind specifically to the combining site of the target antibody, to some other site on the antibody, to some other component of the immobilization matrix, or may simply be non-specifically bound to the components present. A phage lift (*Protocol 10*) probed with the target antibody allows a survey of a great many recovered clones to assess the proportion of those recovered that bind to the antibody. A dot blot of phage from several dozen clones may be probed in the same way and, because of the larger number of phage immobilized, is more sensitive than the colony lift. An aliquot of the enriched pool may be adsorbed to the immobilized antibody under the same conditions used for panning, and the eluted phage titred and compared with the recovery from an irrelevant antibody. The phage ELISA described in *Protocol 9* is used to analyse the specificity of the recovered phage for the antibody. Each clone is amplified and the harvested phage added to wells containing immobilized antibody, the wells are washed free of unbound phage, and the bound phage probed with rabbit anti-phage antisera, which is then detected with anti-rabbit IgG second antibody conjugated to an enzyme such as alkaline phosphatase. This assay is done with and without the target antibody, and can also be done in the presence of a large excess of competing ligand to test for binding of the phage to the combining site of the antibody (15).

Active peptide sequences are deduced by sequencing the DNA in the oligonucleotide cloning site of individual phage clones by standard single-strand dideoxy sequencing methods (*Protocol 11*).

Protocol 8. Phage titreing

Reagents

- *E. coli* K91 (*Protocol 7*)
- LB medium (*Protocol 7*)
- LB-tet agar plates (*Protocol 7*)
- PBS (*Protocol 4*)

Method

1. Inoculate 5 ml of LB medium with a single colony of *E. coli* K91. Grow at 37 °C with shaking overnight.

2. Add 1 ml of the K91 overnight culture to a 100 ml flask containing 20 ml of LB medium and grow with shaking to $A_{600} = 0.2–0.4$. Pellet cells by centrifugation and resuspend in 2 ml LB medium.

3. Prepare ten-fold serial dilutions of phage stock in PBS. Combine 50 µl of diluted phage with 100 µl of concentrated K91 in sterile microfuge tubes. Incubate at 37 °C for 20 min.

4. Plate the infected cells on LB-tet agar plates and incubate at 37 °C overnight.

5. Count the number of tetracycline resistant colonies present on the plates and calculate the total number of phage per ml. The titre of the phage stock is expressed as TU/ml.

Protocol 9. Phage ELISA

Equipment and reagents

- ELISA microplates (Beckman or Dynatech Immulon-4)
- PBS (*Protocol 4*)
- BSA
- Rabbit anti-phage fd antisera (Pharmacia, 5-Prime-3-Prime)
- Goat anti-rabbit IgG–alkaline phosphatase conjugate (Tago)
- Substrate solution (immediately before use add 1 mg/ml *p*-nitrophenyl phosphate to 10% diethanolamine, pH 9.8, 0.5 mM $MgCl_2$)
- 1 M NaOH

Protocol 9. *Continued*
Method

1. Dilute the target antibody to 20 µg/ml in PBS. (For a particular antibody, the amount per well may require adjustment to optimize the
   assay.) Add 50 µl to the appropriate number of wells of a 96-well
   microplate and incubate for 1 h at 37 °C.

2. Wash the plate with PBS. Completely fill the wells with PBS/1.0% BSA
   and incubate for 1 h at 37 °C to block the plate.

3. Wash the wells with PBS. Add 50 µl of phage supernatant and 50 µl of
   PBS/0.1% BSA per well. (If purified phage stocks are used, add (2–5) ×
   10^9 TU per well.) Incubate for 2 h at 4 °C.

4. Wash the wells with PBS and add 50 µl of rabbit anti-phage antisera,
   diluted 1:5000 in PBS/0.1% BSA (proper dilutions must be determined
   for each batch of anti-phage antibody). Incubate for 1 h at 4 °C.

5. Wash the wells with PBS and add 50 µl of alkaline phosphatase-
   conjugated goat anti-rabbit IgG, diluted 1:2000 (or as per supplier's
   instructions) in PBS/0.1% BSA. Incubate for 1 h at 4 °C.

6. Wash the plate with PBS and add 100 µl of substrate solution. After
   development, add 100 µl 1 M NaOH to stop the reaction; measure the
   absorbance at 405 nm with a plate reader.

Protocol 10. Phage lifts and dot blots

Equipment and reagents

- PBS (*Protocol 4*)
- PBST (PBS, 0.05% Tween 20)
- BSA
- Secondary antibody conjugated to alkaline phosphatase (Gibco-BRL)
- Substrate solution (immediately before use

add 44 µl of nitroblue tetrazolium chloride (75 mg/ml in 70% dimethylformamide) and 33 µl of 5-bromo-4-chloro-3-indolyl-phosphate *p*-toluidine salt (50 mg/ml in dimethylformamide) to 10 ml of 0.1 M Tris–HCl, pH 9.5, 0.1 M NaCl, 50 mM MgCl$_2$)
- Vacuum dotting apparatus

Method

1. Titre phage to obtain approximately 200 colonies per plate (*Protocol 8*).

2. *For a phage lift*, place a nitrocellulose circle on the plate, mark the
   position of the filter with a needle, and carefully remove it after 1 min.
   Immediately wash it with PBS to remove bacterial colonies that are
   adhering to the filter. Return the titre plate to the 37 °C incubator for
   several hours to allow colonies to regenerate.

3. *For a phage dot blot*, spot 50–100 µl of phage supernatant (10^9–10^{10}
   TU of a precipitated phage stock) on nitrocellulose using a vacuum
   dotting apparatus.

4. Block the filter in PBST/1% BSA at room temperature for 1 h.

5. Wash the blocked filter with PBST and add primary (target) antibody diluted to 1 µg/ml in PBST/0.1% BSA. Incubate for 1 h at 4°C on a rotating platform.

6. Wash the filter three times with PBST (1 min each wash) and then add alkaline phosphatase-conjugated secondary antibody (diluted 1:1000 in PBST/0.1% BSA). Incubate for 1 h at 4°C with shaking.

7. Repeat the wash with PBST and blot filter dry.

8. Transfer the filter to a new tray and add substrate solution. Incubate at room temperature until colour develops and stop the reaction by washing the filter in water.

Protocol 11. DNA sequencing of phage clones

Reagents

- Sequenase 2.0 (United States Biochemicals)
- pIII sequencing primer (5'-CGA TCT AAA GTT TTG TCG TCT -3')
- PEG/NaCl (*Protocol 4*)
- TE buffer (*Protocol 3*)
- 6% polyacrylamide/7 M urea/TBE gel (see (18))
- Phenol: chloroform (1:1)
- 3 M Sodium acetate

A. *Single-stranded DNA preparation* [a]

1. Add 1 ml of an overnight culture of phage-infected cells to a microfuge tube and spin for 1 min.

2. Transfer the cleared supernatant to a new tube and precipitate phage by adding 0.2 volumes of PEG/NaCl. Mix well and let stand on ice for 1 h.

3. Pellet phage in a microcentrifuge for 5 min at room temperature. Carefully remove as much of the supernatant as possible and resuspend the phage pellet in 100 µl of TE buffer.

4. Extract once with an equal volume of phenol saturated with TE butter. Centrifuge for 1 min and transfer the upper, aqueous phase to a new tube.

5. Extract once with an equal volume of phenol/chloroform. Centrifuge for 1 min and transfer the upper, aqueous phase to a new tube.

6. Precipitate single-stranded phage DNA by adding 0.1 volume 3 M sodium acetate and 2 volumes ethanol. Wash with 70% ethanol, dry the pellet and resuspend in 40 µl TE buffer.

B. *Sequencing*

1. Sequence the single-stranded phage DNA using the dideoxy method and Sequenase 2.0. The pIII sequencing primer anneals 40 base pairs downstream of the 3' *Bst*XI cloning site in fAFF1.

Protocol 11. *Continued*

2. Prepare a 6% polyacrylamide/7 M urea/TBE gel (38 × 50 cm). Load the sequencing reactions and run the gel until the bromophenol blue has migrated off the bottom. (A detailed description of DNA sequencing is found in(16), pp. 13.42–13.69.)

ª A variety of commercial DNA isolation kits are available for the isolation of single-stranded templates for DNA sequencing.

9. Concluding remarks

The availability of very large libraries of random peptides, each peptide linked to its encoding DNA, provides a convenient format for the selection of ligands by a process requiring no knowledge of the structure of the binding site, nor any preconception of the structural requirements of the ligands. To date, the most prominent use of these peptide display techniques has been the mapping of antibody epitopes, the application which we have focused upon in this chapter. There are many other applications where these libraries may prove useful. The identification of novel ligands to receptors, adhesion molecules, enzymes, and a variety of other binding proteins is possible with this approach, and there is rapidly growing interest in the employment of such evolutionary approaches to ligand discovery in many fields of biological research.

Acknowledgements

We would like to acknowledge our many colleagues at Affymax for their contributions toward the development of these protocols. In particular, we thank Ron Barrett, Elizabeth Peters, and Peter Schatz for providing us with unpublished information for inclusion in this chapter. We also thank Ron Barrett, Russell Howard, and Gordon Ringold for their comments on the manuscript.

References

1. Young, R. A., Bloom, B. R., Grosskinsky, C. M., Ivanyi, J., Thomas, D., and Davis, R. W. (1985). *Proc. Natl. Acad. Sci. USA*, **82**, 2583.
2. Smith, G. P. (1985). *Science*, **228**, 1315.
3. Parmley, S. F. and Smith, G. P. (1988). *Gene*, **73**, 305.
4. Parmley, S. F. and Smith, G. P. (1989). *Adv. Exp. Med.*, **251**, 215.
5. Dower, W. J., Miller, J. F., and Ragsdale, C. W. (1988). *Nucleic Acids Res.*, **16**, 6127.
6. Scott, J. K. and Smith, G. P. (1990). *Science*, **249**, 386.

7. Cwirla, S. E., Peters, E. A., Barrett, R. W., and Dower, W. J. (1990). *Proc. Natl. Acad. Sci. USA*, **87**, 6378.
8. Devlin, J. J., Panganiban, L. C., and Devlin, P. E. (1990). *Science*, **249**, 404.
9. Model, P. and Russel, M. (1988). In *The bacteriophages* (ed. R. Calendar), Vol. 2, pp. 375–456. Plenum Press, New York.
10. Zacher, A. N., III, Stock, C. A., Golden, J. W., II, and Smith, G. P. (1980). *Gene*, **9**, 127.
11. Greenwood, J., Willis, A. E., and Perham, R. N. (1991). *J. Mol. Biol.*, **220**, 821.
12. Felici, F., Castagnoli, L., Musacchio, A., Jappelli, R., and Cesareni, G. (1991). *J. Mol. Biol.*, **222**, 301.
13. Cull, M. G., Miller, J. F., and Schatz, P. J. (1992). *Proc. Natl. Acad. Sci. USA*, **89**, 1865.
14. Dower, W. J. and Cwirla, S. E. (1992). In *Guide to electroporation and electrofusion* (ed. D. Chang, B. Chassy, J. Saunders, and A. Sowers), pp. 291–301. Academic Press, San Diego.
15. Barrett, R. W., Cwirla, S. E., Ackerman, M. S., Olson, A. M., Peters, E. A., and Dower, W. J. (1992). *Anal. Biochem.*, **204**, 357.
16. Sambrook, J., Fritsch, E. F., and Maniatis, T. (ed.) (1989). *Molecular cloning: a laboratory manual*. Cold Spring Harbor Press, Cold Spring Harbor, New York.
17. Arufo, A. and Seed, B. (1987). *Proc. Natl. Acad. Sci. USA*, **84,** 8573.
18. Ausubel, F. M., Brent, R., Kingston, R. E., Moore, D. D., Seidman, J. G., Smith, J. A., and Struhl, K. (ed.) (1987). *Current protocols in molecular biology*. John Wiley, New York.

A1

Suppliers of specialist items

Affymax Research Institute, 4001 Miranda Avenue, Palo Alto, CA 94304, USA.

Aldrich Chemical, The Old Brickyard, New Road, Gillingham, Dorset SP8 4JL, UK; 1001 West Saint Paul Avenue, Milwaukee, WI 53233, USA.

American Type Culture Collection, 12301 Parklawn Drive, Rockville, MD 20852–1776, USA.

Amersham International, Lincoln Place, Green End, Aylesbury, Bucks HP20 2TP, UK; 2636 South Clearbrook Drive, Arlington Heights, IL 60005, USA.

Amicon, Upper Mill, Stonehouse, Glos GL10 2BJ, UK; 17 Cherry Hill Drive, Danvers, MA 01923, USA.

Applied Biosystems, 850 Lincoln Centre Drive, Foster City, CA 94404, USA; Kelvin Close, Birchwood Science Park, Warrington, Cheshire WA3 7PB, UK.

Bachem Feinchemikalien, Haupstrasse 144, CH-4416 Bubedorf, Switzerland.

J. T. Baker, Rijsterborgherweg 20, PO Box 1, 7400 AA Deventer, The Netherlands.

BDH, Broom Road, Poole, Dorset BH12 4NN, UK.

Beckman Instruments, 6200 El Camino Real, Carlsbad, CA 92008, USA.

BioProbe International, 14272 Franklin Avenue, Tustin, CA 92680, USA.

Bioprocessing Ltd, Medomsley Road, Consett, Co. Durham DH8 6TJ, UK.

Bio-Products, B3 Bersham Enterprise Centre, Plas Grono Road, Rhostyllen, Wrexham, Clwyd LL14 4EG, UK.

Bio-Rad Laboratories, Bio-Rad House, Maylands Road, Hemel Hempstead, Herts HP2 7TD, UK; 3300 Regatta Road, Richmond, CA 94804, USA.

Biotech Instruments, 75A High Street, Kimpton, Herts SG4 8PU, UK.

Boehringer Mannheim, Bell Lane, Lewes, East Sussex BN7 1LG, UK; 9115 Hague Road, PO Box 50414, Indianapolis, IN 46250–0414, USA.

Calbiochem-Novabiochem, 3 Heathcoat Building, Highfields Science Park, University Boulevard, Nottingham NG7 2QJ, UK; PO Box 12087, San Diego, CA 92112–4180, USA.

Cambridge Research Biochemicals, Gadbrook Park, Northwich, Cheshire CW9 7RA, UK; Wilmington, DE 19897, USA.

Chiron Mimotopes, 11 Duerdin Street, Clayton, Victoria 3168, Australia.

Clontech Laboratories, 4030 Fabian Way, Palo Alto, CA 94303–4607, USA.

CPG, 32 Pier Lane West, Fairfield, NJ 07006, USA.

Dako, Produktionsvej 42, PO Box 1359, DK 2600 Glostrop, Denmark; 6392 Via Real, Carpinteria, CA 93013, USA.

Dupont, Corcord Plaza-Qillen Bldg Wilmington, DE 19898, USA; Wedywood Way, Stevenage, Herts, UK.

Dynatech, 3 New England Executive Park, Burlington, MA 01803, USA; Daux Road, Billingshurst, West Sussex RH14 9SJ, UK.

Flow Laboratories, 7655 Old Springhouse Road, McClean, VA 22102, USA; Woodcock Hill, Harefield Road, Rickmansworth, Herts WD3 1PQ, UK.

Gibco-BRL, Trident House, PO Box 35, Renfrew Road, Paisley, Renfrewshire PA3 4EF, UK; 8400 Helgerman Court, Gaithersburg, MD 20877, USA.

Heffers, 19 Sidney Street, Cambridge CB2 3HL, UK.

IBF Biotechnics, 35 avenue Jean Jaures, 92390 Villeneuve la Garenne, France; 7151 Columbia Gateway Drive, Columbia, MD 21046, USA.

Immunodiagnostic Systems, Balden Business Park, Balden, Tyne and Wear NE35 9PD, UK.

Kem-En-Tec, Haraldsgade 68, DK-2100 Copenhagen, Denmark.

Labsystems Group, Unit 5, The Ringway Centre, Edison Road, Basingstoke, Hants RG21 2YH, UK.

LDC, Diamond Way, Stone Business Centre, Stone, Staffs ST15 0HH, UK.

Luckham, Victoria Gardens, Burgess Hill, West Sussex RH15 9QN, UK.

MilliGen/Biosearch, see Millipore.

Millipore, 186 Middlesex Turnpike, Burlington, MA 01803, USA; The Boulevard, Blackmoor Lane, Watford, Herts WD1 8YW, UK.

New England Biolabs, 32 Tozer Road, Beverly, MA 11915–5599, USA; 67 Knowl Piece, Wilbury Way, Hitchin, Herts SG4 0TY, UK.

Novabiochem, see Cabiochem-Novabiochem.

Nunc, Postbox 280, Kamstrup, DK4000 Roskilde, Denmark.

Omnifit, 51 Norfolk Street, Cambridge CB1 2LE, UK.

Peninsula Laboratories, 611 Taylor Way, Belmont, CA 94002, USA; Box 62, 17K Westside Industrial Estate, Jackson Way, St Helens, Merseyside WA9 3AJ, UK.

Peptides International, PO Box 24658, Louisville, KY 40224, USA.

PerSpective Biosystems, University Park at MIT, 38 Sidney Street, Cambridge, MA 02139, USA.

Pharmacia LKB Biotechnology, Davy Avenue, Knowhill, Milton Keynes MK5 8PH, UK; 800 Centennial Avenue, PO Box 1327, NJ 08855–1327, USA.

Pierce, 3747 North Meridian Road, PO Box 117, Rockford, IL 61105, USA; PO Box 1512, 3260 BA Oud-Beijerland, The Netherlands.

5-Prime-3-Prime, 5603 Araphoe Road, Boulder, CO 80303, USA.

Rocky Mountain Scientific Glassblowing, Denver, CO, USA.

Serotec, 22 Bankside, Station Approach, Kidlington, Oxford OX5 1JE, UK.

Sigma Chemical, Fancy Road, Poole, Dorset BH17 7NH, UK; 3050 Spruce Street, St Louis, MO 63103, USA.

Tago, PO Box 4463, 887 Mitten Road, Burlingame, CA 94011, USA.

TosoHaas, Rohm & Hass Bldg, Independence Mall West, Philadelphia, PA 01757, USA.

United States Biochemicals, PO Box 22400, Cleveland, OH 44122, USA.

Vector Laboratories, 30 Ingold Road, Buringame, CA 94101, USA; 16 Wulfric Square, Bretton, Peterborough PE3 8RF, UK.

Vilber Lourmat, BP66, 77202 Marne la Vallee, Cedex 1, France.

Whatman Scientific, St Leonard's Road, 20/20 Maidstone, Kent ME 16 0LS, UK.

The DNA-encoded amino acids[a]

Amino acid	Symbols		Mol. wt.	pK values			pI
Alanine	Ala	A	89	2.35	9.87		6.11
Arginine	Arg	R	174	2.01	9.04	12.48	10.76
Asparagine[b]	Asn	N	132	2.14	8.72		5.43
Aspartic acid[b]	Asp	D	133	2.10	3.86	9.82	2.98
Cysteine	Cys	C	121	1.92	8.37	10.70	5.15
Glutamic acid[c]	Glu	E	147	2.10	9.47	4.07	3.08
Glutamine[c]	Gln	Q	146	2.17	9.13		5.65
Glycine	Gly	G	75	2.35	9.78		6.06
Histidine	His	H	155	1.77	9.18	6.10	7.64
Isoleucine	Ile	I	131	2.32	9.76		6.04
Leucine	Leu	L	131	2.33	9.74		6.04
Lysine	Lys	K	146	2.18	8.95	10.53	9.47
Methionine	Met	M	149	2.13	9.28		5.71
Phenylalanine	Phe	F	165	2.20	9.31		5.76
Proline	Pro	P	115	2.00	10.60		6.30
Serine	Ser	S	105	2.19	9.21		5.70
Threonine	Thr	T	119	2.09	9.10		5.60
Tryptophan	Trp	W	204	2.38	9.39		5.88
Tyrosine	Tyr	Y	181	2.20	9.11	10.07	5.63
Valine	Val	V	117	2.29	9.74		6.02

[a] Further data are available in Fasman, G. D. (1976). *Handbook of biochemistry and molecular biology*, 3rd edn. CRC Press, Cleveland, OH; and in Dawson, R. M. C., Elliott, D. C., Elliott, W. H., and Jones, K. M. (1986). *Data for biochemical research*, 3rd edn. Clarendon Press, Oxford.

[b] When it not possible to distinguish asparate and asparagine the abbreviation Asx or B is used.

[c] When it not possible to distinguish glutamine and glutamate the abbreviation Glx or Z is used.

Index